Modeling *Our* World

The ESRI® Guide to Geodatabase Design

Michael Zeiler

ESRI Press

PUBLISHED BY
Environmental Systems Research Institute, Inc.
380 New York Street
Redlands, California 92373-8100

Environmental Systems Research Institute, Inc.
Modeling Our World
The ESRI Guide to Geodatabase Design
ISBN 1-879102-62-5

Contents

Preface

All geographic information systems (GIS) are built using formal models that describe how things are located in space. A formal model is an abstract and well-defined system of concepts. It defines the vocabulary that we can use to describe and reason about things. A geographic data model defines the vocabulary for describing and reasoning about the things that are located on the earth. Geographic data models serve as the foundation on which all geographic information systems are built.

We are all familiar with one model for geographic information—the map. A map is a scale model of reality that we build, using a set of conventions and rules (for example, map projections, line symbols, text). Once we construct a map, we can use it to answer questions about the reality it represents. For example, how far is it from Los Angeles to San Diego? Or, what cities lie along the Mississippi River? The map model also serves as a tool for communicating facts about geography visually: Is the terrain rough? Which way is north? In fact, when we *see* a map, we often understand things that might not even occur to us as specific questions.

Maps work because we know the "rules" of conventional map reading: blue lines are rivers, North is toward the top of the page, and so on. In a similar way, geographic data models define their own set of concepts and relationships, which must be understood before you can expect to create or interpret your own data model. These concepts relate to how you can represent geographic information in a computer system, rather than, as in the map example, on paper.

In *Modeling Our World,* Michael Zeiler has written an excellent primer for understanding the various models used to represent geographic information in ArcInfo™ 8 software. He presents, using straightforward text and excellent illustrations, the concepts and vocabulary employed in the design, implementation, and use of the ArcInfo 8 geographic database. In addition to explaining the ArcInfo data model (objects, features, surfaces, networks, images, and so forth) in detail, Michael also provides good insight into how to use this framework to design useful information models that fit your particular needs.

This book serves a variety of different purposes. For the geographer or scientist, it defines a conceptual context for representing geographic information. For the GIS specialist, it serves as a guidebook in designing and using geographic databases. Finally, it introduces database concepts to a geographic audience, and geographic concepts to the database specialist.

ArcInfo 8 defines a unified framework for representing geographic information in a database. Several different generic data models are supported within this framework:

- cell-based or raster representation

- object-based or feature-based representation

- network or graph-element representation

- finite-element or TIN representation

Each of these generic models has its own vocabulary used to define and reason about geographic information. When we decide to represent roads, rivers, terrain, or any sort of phenomena in a GIS, we need to decide exactly how we define information in terms of these generic models. As chapter 1 points out, there are many ways that information can be modeled in a GIS. The representation you choose for the data model will affect how you sample and measure geographic information, how you display it visually, and which relationships between elements can be represented, as well as query and analysis operations that can be applied to the information.

Some have asserted that we should hide representational models for geographic information (features, geometry, rasters, surfaces, and so on) from the users of geographic information systems. Somehow, these representational concepts are considered "implementation details." In this view, a single real-world thing, such as the Mississippi River, should be modeled as a single thing within the GIS. Perhaps, behind the scenes, the system could automatically use multiple representations for these real-world things. If you ask "What is upstream?" it could use a network representation of the river. If you ask "What is the surface area of the water?" it could use a polygon feature representation. If you ask "What area does it drain?" it could use a surface or terrain representation, and so on. While it may be desirable to hide these concepts from some consumers of geographic information, I believe that a strong understanding of geographic data models and representations is crucial to the correct design and use of geographic information systems. Geographic data models act as the lens or filter through which we perceive and interpret the infinite complexity of the real world. It is only in the context of representations of the Mississippi River, for example, that we can define specific properties, behavior, or even its identity as a "thing of interest." Understanding geographic data model concepts is central to knowing how to define and collect geographic information. It is also crucial for correctly interpreting the results derived from the analysis of geographic information. This is similar to the role that statistics and sampling theory play in the natural sciences.

For the GIS specialist, this book serves as an introduction to a new object-relational model for representing features, spatial relationships between features, and other thematic relationships. This new model is significantly richer in its ability to represent features with associated behavior, relationships, and properties than the current coverage or shapefile model. If you are already familiar with coverages, shapefiles, and database tables, the new model is a dramatic extension of concepts and capabilities with which you are already familiar. Our goal in building the new feature data model has been to move as much specialized application logic (for example, maintaining connectivity or relational integrity between objects) as possible into the scope of the data model itself. This allows more of the GIS application to be defined using rules in the data

model, rather than custom application logic written for each application. For other aspects of the data model, which may already be familiar to the reader, the specific jargon and concepts used in ArcInfo 8 (for topics like image data, as an example) are clearly introduced and defined.

This book also connects the specialized world of geographic information systems and the broader world of object-relational databases. ArcInfo now supports the direct use of standard relational database technology as an integral part of the GIS. This introduces some new concepts to the GIS community. Topics such as transaction models for simultaneous editing of a shared, seamless database are described in detail. For the GIS specialist, this provides a good introduction to standard database concepts. For the database specialist, this book serves as a good answer to the question "what is so special about spatial?"

Working with geographic information systems is fun for me because it serves to integrate concepts and ideas from a variety of different disciplines— geometry and networks from applied mathematics, sampling and measurement theory from remote sensing and physics, information modeling and multiuser database issues from information technology. In working with GIS, we get to integrate all of this in a single, useful framework for building real systems. This book presents that synthesis, based on our work with ArcInfo 8. I hope you find this book useful and stimulating as a basis for your own work in geographic information systems.

Scott Morehouse
Director of Software Development
Environmental Systems Research Institute, Inc.
Redlands, California

Acknowledgments

This book, *Modeling Our World,* is the distillation of many people's inspirations, ideas, and labors.

Many deserve recognition—the ArcInfo user community, which always amazes us with creative applications of GIS; the ArcInfo 8 development team, which has produced a true masterpiece of software; and the teams throughout ESRI, which collaborated to take GIS technology to new levels.

Because of the constraints of space, only a few can be directly acknowledged. These are some of the contributors to this software release and book.

The structural design of ArcInfo 8 was led by some of the brightest thinkers in the industry. Sud Menon directed the architectural design of the geodatabase and he is responsible for many of the insights expressed in this book. Jeff Jackson led the implementation of software component technology that has revolutionized ArcInfo. Erik Hoel applied his expertise to the development of the network features and the framework for vertical applications. The development of the ArcMap™ and ArcCatalog™ applications was led by Barry Michaels, Scott Simon, and Keith Ludwig. The accessibility and consistency of the software user interface was guided by Rupert Essinger. This complex endeavor was orchestrated by Matt McGrath.

Many product specialists and programmers at ESRI provided material for this book and reviewed chapters. These include Andy MacDonald, Charlie Frye, Mike Minami, Aleta Vienneau, Jim TenBrink, Wolfgang Bitterlich, Tom Brown, Dale Honeycutt, Steve Kopp, Brett Borup, Peter Petri, Clayton Crawford, and Andrew Perencsik. The contributions of Andy, Dale, and Steve to chapters 5, 8, and 9 respectively are particularly noteworthy.

The attractive city maps throughout this book were kindly provided by Gar Clarke, GIS manager at the City of Santa Fe, New Mexico. The image of Mars at the front of chapter 9 is courtesy of Malin Space Science Systems and JPL/NASA.

The maps on the chapter title pages are drawn from the work of many cartographers from history. Their maps remind us that, although we have reached a level of sophistication in drawing maps with computers, we have yet to equal their artistry.

Several people were actively engaged in the production of this book. Jennifer Wrightsell rigorously edited the chapters and designed the layout, along with Andy Mitchell and Youngiee Auh. Amaree Israngkura designed the cover. Michael Hyatt did the copyedit. Robin Floyd and Christian Harder managed and guided the publication of this book.

Scott Morehouse wrote the preface and is ESRI's visionary on advancing the theory and practice of GIS. Clint Brown prodded and inspired us to create the best product we had within ourselves. Curt Wilkinson and David Maguire worked hard to ensure that ArcInfo 8 meets the goals and requirements of users. Jack Dangermond created this very special and unique institute where we can believe that we make a difference in this world and act on that idea.

Finally, my wife Elizabeth deserves special thanks for her countless hours of support. Her commitment and encouragement made the effort to produce this book possible.

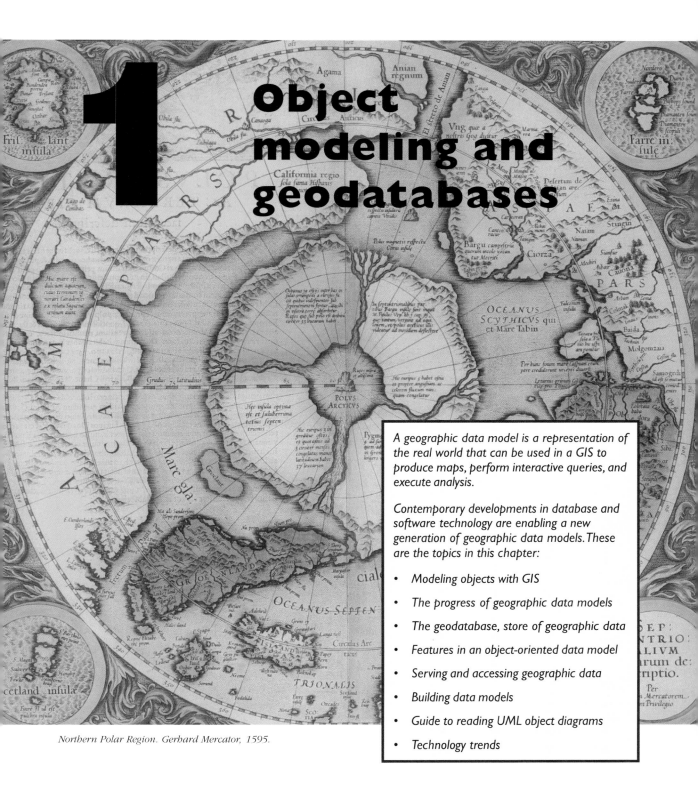

1 Object modeling and geodatabases

A geographic data model is a representation of the real world that can be used in a GIS to produce maps, perform interactive queries, and execute analysis.

Contemporary developments in database and software technology are enabling a new generation of geographic data models. These are the topics in this chapter:

- Modeling objects with GIS

- The progress of geographic data models

- The geodatabase, store of geographic data

- Features in an object-oriented data model

- Serving and accessing geographic data

- Building data models

- Guide to reading UML object diagrams

- Technology trends

Northern Polar Region. Gerhard Mercator, 1595.

The purpose of a geographic information system (GIS) is to provide a spatial framework to support decisions for the intelligent use of earth's resources and to manage the man-made environment.

Most often, a GIS presents information in the form of maps and symbols. Looking at a map gives you the knowledge of where things are, what they are, how they can be reached by means of roads or other transport, and what things are adjacent and nearby. A GIS can also disseminate information through an interactive session with maps on a personal computer. This interaction reveals information that is not apparent on a printed map.

For example, you can query all known attributes of a feature, create a list of all things connected from one point on a network to another, and perform simulations to gauge qualities such as water flow, travel time, or dispersion of pollutants.

The way you choose to display and analyze information depends upon how you model geographic objects from the world.

MANY WAYS TO MODEL A SYSTEM

Our interaction with objects in the world is diverse, and you can model them in many ways.

Consider one example, rivers. Rivers are natural features, are used for transportation, delimit political or administrative areas, and are an important feature in the shape of a surface. Here are a few of the many ways you can think about modeling rivers in a GIS:

- As a set of lines that form a network. Each section of line has flow direction, volume, and other attributes of a river. You can apply a linear network model to analyze hydrographic flow or ship traffic.

- As a border between two areas. A river can delimit political areas such as provinces or counties, or can be a barrier for natural regions such as wildlife habitats.

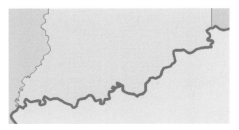

- As an areal feature with an accurate representation of its banks, braids, and navigable channels on the river.

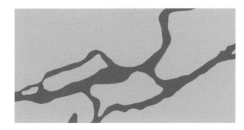

- As a sinuous line forming a trough in a surface model. From the river's path through a surface, you can calculate its profile and rate of descent, the watershed it drains, and its flooding potential for a prescribed rainfall.

MAP USE GUIDES THE DATA MODEL

It is clear that even a common type of geographic feature such as a river can be represented in a GIS in a variety of ways. No model is intrinsically superior; the type of map you want to create and the context of the problems to be solved will guide which model is best.

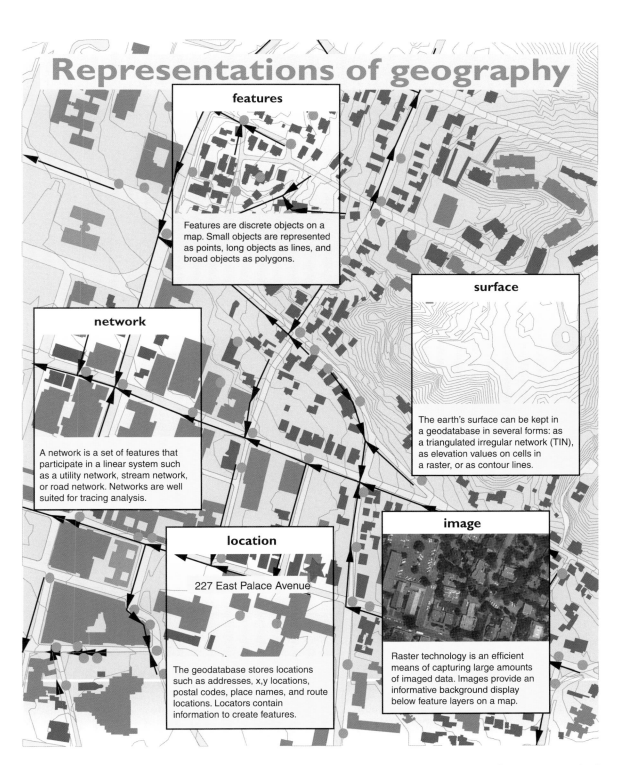

Representations of geography

features

Features are discrete objects on a map. Small objects are represented as points, long objects as lines, and broad objects as polygons.

network

A network is a set of features that participate in a linear system such as a utility network, stream network, or road network. Networks are well suited for tracing analysis.

surface

The earth's surface can be kept in a geodatabase in several forms: as a triangulated irregular network (TIN), as elevation values on cells in a raster, or as contour lines.

location

227 East Palace Avenue

The geodatabase stores locations such as addresses, x,y locations, postal codes, place names, and route locations. Locators contain information to create features.

image

Raster technology is an efficient means of capturing large amounts of imaged data. Images provide an informative background display below feature layers on a map.

A geographic data model is an abstraction of the real world that employs a set of data objects that support map display, query, editing, and analysis.

ArcInfo 8 introduces a new object-oriented data model—the geodatabase data model—that is capable of representing natural behaviors and relationships of features. To understand the impact of this new model, it is instructive to review three generations of geographic data models.

THE CAD DATA MODEL

The very first computerized mapping systems drew vector maps with lines displayed on cathode ray tubes and raster maps using overprinted characters on line printers. From this genesis, the 1960s and 1970s saw the refinement of graphics hardware and mapping software that could render maps with reasonable cartographic fidelity.

In this era, maps were usually created with general-purpose CAD (computer-aided design) software. The *CAD data model* stored geographic data in binary file formats with representations for points, lines, and areas. Scant information about attributes was kept in these files; map layers and annotation labels were the primary representation of attributes.

THE COVERAGE DATA MODEL

In 1981, Environmental Systems Research Institute, Inc. (ESRI), introduced its first commercial GIS software, ArcInfo, which implemented a second-generation geographic data model, the *coverage data model* (also known as the georelational data model). This model has two key facets:

- Spatial data is combined with attribute data. The spatial data is stored in indexed binary files, which are optimized for display and access. The attribute data is stored in tables with a number of rows equal to the number of features in the binary tables and joined by a common identifier.

- Topological relationships between vector features can be stored. This means that the spatial data record for a line contains information about which nodes delimit that line, and by inference, which lines are connected; it also contains information about which polygons are on its right and left sides.

The major advance of the coverage data model was the user's ability to customize feature tables; not only could fields be added, but database relates could be set up to external database tables.

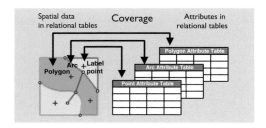

Because of the performance limitations of computer hardware and database software of the time, it was not practical to store spatial data directly in a relational database. Rather, the coverage data model combined spatial data in indexed binary files with attribute data in tables.

Despite this compromise of partitioning spatial and attribute data, the coverage data model has become the dominant data model in GIS. This has been for good reason—the coverage data model made high-performance GIS possible, and stored topology facilitated improved geographic analysis and more accurate data entry.

Limitations of the coverage data model

However, the coverage data model has an important shortcoming—features are aggregated into homogeneous collections of points, lines, and polygons with generic behavior. The behavior of a line representing a road is identical to the behavior of a line representing a stream.

The generic behavior supported by the coverage data model enforces the topological integrity of a dataset. For example, if you add a line across a polygon, it is automatically split into two polygons.

But it is desirable to also support the special behaviors of streams, roads, and other real-world objects. An example is that streams flow downhill and when two streams merge into one, the flow of the merged stream is the addition of the two upstream flows. Another example is that when two roads cross, a traffic intersection should be at their junction unless there is an overpass or underpass.

Customizing features in coverages

With the coverage data model, ArcInfo application developers had some notable success in adding this type of behavior to features through macro code written in the ARC Macro Language (AML™). Many successful, large-scale, industry-specific applications were built.

However, as applications became more complex, it became apparent that a better way to associate behavior with features was needed. The problem was that the developer had the task of keeping the application code in synchronicity with feature classes—no easy task. The time had come for a new geographic data model with an infrastructure to tightly couple behavior with features.

THE GEODATABASE DATA MODEL

ArcInfo 8 introduces a new object-oriented data model called the *geodatabase data model*. The defining purpose of this new data model is to let you make the features in your GIS datasets smarter by endowing them with natural behaviors, and to allow any sort of relationship to be defined among features.

The geodatabase data model brings a physical data model closer to its logical data model. The data objects in a geodatabase are mostly the same objects you would define in a logical data model, such as owners, buildings, parcels, and roads.

Further, the geodatabase data model lets you implement the majority of custom behaviors without writing any code. Most behaviors are implemented through domains, validation rules, and other functions of the framework provided in ArcInfo. Writing software code is only necessary for the more specialized behaviors of features.

SCENARIOS OF OBJECT INTERACTIONS

To get a sense of why an object-oriented data model is important, review the following scenarios that illustrate common tasks you might perform with features. From these scenarios, you can sift out the benefits of an object-oriented data model and then review some specific characteristics of the geodatabase data model.

Adding and editing features

When you add geographic features to your GIS database, you want to ensure that features are placed correctly according to rules such as these:

- That the values you assign to an attribute fall within a prescribed set of permissible values. A parcel of land may only have certain land uses such as residential, agricultural, or industrial.

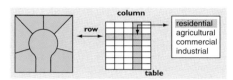

- That a feature can be placed adjacent or connected to another feature only if certain constraints are met. Placing a liquor store near a school is not permitted by law. A city road cannot be connected to a highway without a transition segment such as an on-ramp.

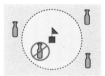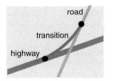

- That collections of certain features conform to their natural spatial arrangement. A stream system should always flow downhill. Flow down from a junction is the sum of flows upstream.

- That the geometry of a feature follows its logical placement. The lines and curves that make up a road should be tangent. Building corners most often form right angles.

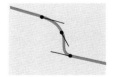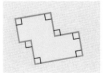

Relationships among features

All objects in the world are entangled in relationships with other objects. From the perspective of a GIS, these relationships can be considered to fall within three general categories: topological, spatial, and general.

These are some examples of each of these types of relationships:

- When you edit features in an electric utility system, you want to be sure that the ends of primary and secondary lines connect exactly and that you are able to perform tracing analysis on that electric network. A set of topological relationships is defined for you when you load or edit features within a connected system.

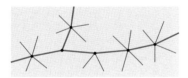

- When you work with a map with buildings, blocks, and school districts, you might want to determine which block contains a particular building, the set of all buildings within a school district, and which blocks contain no buildings. A fundamental function of a GIS is to determine whether a feature is inside, touching, outside, or overlapping another feature. Spatial relationships are inferred from the geometry of features.

- Some objects have relationships that are not present on a map. A parcel has a relationship to an owner, but the owner is not a feature on a map. A general relationship connects the parcel and the owner. Some features on a map have relationships, but their spatial relationship is ambiguous. A utility meter is in the general vicinity of an electric transformer, but it is not touching the transformer. The meter and the transformer might not be reliably related by their spatial proximity in crowded areas, so a general relationship ties the two features together.

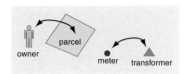

Cartographic display

Most of the time, you will draw features on a map with predefined symbols, but sometimes you will want more control over how your features are drawn. These are some specialized drawing behaviors:

- When you display a contour line, you want its elevation annotated along a flat section of the contour, at an average interval such as 4 inches, and not obscuring other features.

- When you draw roads on a detailed map, you would like the road drawn as parallel lines with clean intersections wherever there is a road intersection.

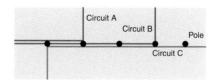

- When multiple electrical wires are physically mounted on the same set of utility poles, you would like to depict them as spread in a set of parallel lines with a standard offset in map units.

Interactive analysis

Dynamic map displays invite the user to touch features, find properties and relationships, and launch analyses. These are examples of some tasks you may want to perform upon selected features:

- Touch a feature on a map display and invoke a form to query and update its properties.

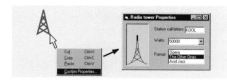

- Select a part of an electric network where line maintenance is planned, find all affected downstream customers, and make a mailing list to notify them.

BENEFITS OF THE GEODATABASE DATA MODEL

The common thread throughout these scenarios is that it is very useful to apply object-oriented data modeling to features. Object-oriented data modeling lets you characterize features more naturally by letting you define your own types of objects, by defining topological, spatial, and general relationships, and by capturing how these objects interact with other objects. Some of the benefits of the geodatabase data model are:

- *A uniform repository of geographic data.* All of your geographic data can be stored and centrally managed in one database.

- *Data entry and editing is more accurate.* Fewer mistakes are made because most of them can be prevented by intelligent validation behavior. For many users, this alone is a compelling reason to adopt the geodatabase data model.

- *Users work with more intuitive data objects.* Properly designed, a geodatabase contains data objects that correspond to the user's model of

data. Instead of generic points, lines, and areas, the users work with objects of interest, such as transformers, roads, and lakes.

- *Features have a richer context.* With topological associations, spatial representation, and general relationships, you not only define a feature's qualities, but its context with other features. This lets you specify what happens to features when a related feature is moved, changed, or deleted. This context also lets you locate and inspect a feature that is related to another.

- *Better maps can be made.* You have more control over how features are drawn and you can add intelligent drawing behavior. You can apply sophisticated drawing methods directly in the ArcInfo mapping application, ArcMap. Highly specialized drawing methods can be executed by writing software code.

- *Features on a map display are dynamic.* When you work with features in ArcInfo, they can respond to changes in neighboring features. You can also associate custom queries or analytic tools with features.

- *Shapes of features are better defined.* The geodatabase data model lets you define the shapes of features using straight lines, circular curves, elliptical curves, and Bézier splines.

- *Sets of features are continuous.* By their design, geodatabases can accommodate very large sets of features without tiles or other spatial partitions.

- *Many users can edit geographic data simultaneously.* The geodatabase data model permits work flows where many people can edit features in a local area, and then reconcile any conflicts that emerge.

To be sure, you can realize some of these benefits without an object-oriented data model, but you would be at a disadvantage—you would need to write external code loosely coupled to features and prone to complexity and error. A principal advantage of the geodatabase data model is that it includes a framework to make it as easy as possible to create intelligent features that mimic the interactions and behaviors of real-world objects.

A geodatabase can contain four representations of geographic data:

- Vector data for representing features

- Raster data for representing images, gridded thematic data, and surfaces

- Triangulated irregular networks (TINs) for representing surfaces

- Addresses and locators for finding a geographic position

A geodatabase stores all of these representations of geographic data in a commercial relational database. This means that geographic data can be administered centrally by information technology professionals and ArcInfo can take advantage of developments in database technology.

REPRESENTING FEATURES WITH VECTORS

Many of the features in the world have well-defined shapes. Vector data represents the shapes of features precisely and compactly as an ordered set of coordinates with associated attributes. This representation supports geometric operations such as calculating length and area, identifying overlaps and intersections, and finding other features that are adjacent or nearby.

Vector data can be classified by dimension:

- Points are zero-dimensional shapes representing geographic features too small to be depicted as lines or areas. Points are stored as a single x,y coordinate with attributes.

- Lines are one-dimensional shapes that represent geographic features too narrow to depict as areas. Lines are stored as a series of ordered x,y coordinates with attributes. The segments of a line can be straight, circular, elliptical, or splined.

- Polygons are two-dimensional shapes that represent broad geographic features stored as a series of segments that enclose an area. These segments form a set of closed areas.

Another type of vector data is annotation. These are descriptive labels that are associated with features and display names and attributes.

Vector data in a geodatabase has a structure that directs the storage of features by their dimension and relationships. A feature dataset is the container of spatial entities (features) and nonspatial entities (objects) and the relationships between them. Topological associations are represented with geometric networks and planar topologies.

A geodatabase also stores validation rules and domains to ensure that when features are created or updated, their attributes remain valid in the context of related features and objects.

REPRESENTING GRIDDED DATA WITH RASTERS

Much of the data collected in a geodatabase is in grid form. This is because cameras and imaging systems record data as pixel values in a two-dimensional grid, or raster.

A cell is a pixel element of a raster and its values can depict a variety of data. A cell can store the reflectance of light for part of the spectrum, a color value for a photograph, a thematic attribute such as vegetative type, a surface value, or elevation.

REPRESENTING SURFACES WITH TINS

A triangulated irregular network (TIN) is a model of a surface. A geodatabase stores TINs as an integrated set of nodes with elevations and triangles with edges. An elevation (or z value) can be interpolated for any point within the geographic extent of a TIN.

TINs enable surface analysis such as watershed studies, visibility of a surface from an observation point, and delineation of surface features such as ridges, streams, and peaks. TINs can also depict the physical relief of terrain.

Note: At the initial release of ArcInfo 8, a geodatabase does not yet store TINs or rasters. For the interim, TINs can be stored in coverage workspaces and rasters in folders or workspaces.

FINDING ADDRESSES WITH LOCATORS

Perhaps the most common geographic task is finding an address. A geodatabase can store addresses and other locations. Geodatabases also store locators containing information that allows you to create features for locations.

Inside a geodatabase

Geodatabase

Feature datasets

```
┌ ─ ─ ─ ─ ─ ─ ─ ─ ─ ─ ─ ─ ─ ┐
│ Spatial reference         │
└ ─ ─ ─ ─ ─ ─ ─ ─ ─ ─ ─ ─ ─ ┘
```

Object classes, subtypes

can be inside or outside of feature datasets

Feature classes, subtypes

Relationship classes

Geometric networks

Planar topologies

Domains

Validation rules

Raster datasets

Rasters

TIN datasets

nodes   edges

faces

Locators

Addresses
x,y locations
ZIP Codes
Place names
Route locations

All feature classes in a feature dataset share a common coordinate system. Because the feature dataset is the container of topological associations, it is important to guarantee a common spatial reference.

A feature dataset contains objects and features and the relationships among them. An object is a nonspatial entity and a feature is a spatial entity. A relationship links two entities.

Objects of the same kind are stored in an object class. Features of the same kind and with the same type of geometric shape are stored in a feature class.

A relationship class stores relationships between entities in two object or feature classes.

Geometric networks model linear systems such as utility networks and transportation networks. They support a rich set of network-tracing and -solving functions.

Planar topologies model systems of line and area features as a continuous coverage of an area. Planar topologies allow features to share common boundaries, such as counties sharing an outer boundary with a state.

Domains are sets of valid attribute values for object attributes. They can be textual or numeric.

Validation rules enforce data integrity through relationship rules and connectivity rules.

Raster datasets can represent an imaged map, a surface, an environmental attribute sampled on a grid, or photographs of objects referenced to features. Some raster data is collected in bands that commonly represent different spectral ranges of camera filters.

TIN datasets are triangulations of sets of irregularly located points with z-values (elevations) sampled from a surface. TINs are most often used to model the earth's surface, but are also used to study the distribution of a continuous environmental factor such as chemical concentration.

Corporate and agency databases have many records with addresses and other locations. Locators contain information that allows you to create features for locations so you can display them on a map.

ArcInfo 8 is distinguished from antecedent releases as it applies object-oriented methodology to geographic data modeling. A developer interacts with data objects through a framework of object-oriented software classes called the *geodatabase data access objects*.

There are three key hallmarks of object orientation: polymorphism, encapsulation, and inheritance.

- *Polymorphism* means that the behaviors (or methods) of an object class can adapt to variations of objects. For example, the core behavior of features, such as draw, add, and delete operations, is the same whether the features reside in a geodatabase, coverage, or shapefile.

- *Encapsulation* means that an object is accessed only through a well-defined set of software methods, organized into software interfaces. The geodatabase data access objects mask the internal details of data objects and provide a standard programming interface.

- *Inheritance* means that an object class can be defined to include the behavior of another object class and have additional behaviors. You can create custom feature types in ArcInfo and inherit the behavior of standard features. For example, a transformer object can be extended (or subtyped) from a standard ArcInfo feature type such as a simple junction feature.

UNIFIED DATA MODEL

The geodatabase data access objects comprise software technology that provides uniform access to geographic data from several data sources such as geodatabases, coverages, and shapefiles.

ArcInfo developers interact with geographic data through a set of data items, such as datasets, tables, feature classes, rows, objects, and features. These items comprise a common and consistent view of geographic data.

Because of this unified data model, the ArcInfo user can work with geodatabases, coverages, and shapefiles in the same way. The unified data model simplifies how users work with data by emphasizing the common characteristics of data.

EXTENSIBLE FEATURES

An important aspect of a geodatabase is that you can optionally create custom features such as transformers and roads, instead of points and lines.

To the ArcInfo user, this means that a transformer or road has all of the standard display, query, and edit behavior of standard point features and line features, but with additional behaviors. You can specify that a transformer must be drawn touching a power pole and perpendicular to the electric line through the pole. Or, when a road is edited, all of its segments must be tangent.

A data modeler can use standard feature types to implement a rich data model. For advanced applications, a developer can extend the standard feature types and create custom features using the object-oriented technique of type inheritance.

Any custom feature that you create enjoys the same performance and functionality as the standard feature types provided by ArcInfo. This offers limitless opportunities for sophisticated application development.

FEATURES AND OBJECT ORIENTATION

Features in a geodatabase are implemented as a set of relational tables. Some of these tables represent collections of features. Other tables represent relationships between features, validation rules, and attribute domains.

ArcInfo manages the structure and integrity of these tables and presents an object-oriented geographic data model through the geographic data access objects.

All users and most developers will not know or care about the details of the internal structure of a geodatabase. The ArcCatalog application is your user interface to establish, modify, and refine the structure of your geodatabase.

The object view of data lets you focus your efforts on building a geographic data model and hides most of the physical database structure of the geodatabase.

Features in a geodatabase

unified data model

ArcInfo is versatile at displaying and analyzing geographic features. ArcInfo works with a number of data sources, including geodatabases, coverages, and shapefiles.

The *geodatabase data access objects* comprise a programming interface that largely hides any differences among feature types from geodatabases, coverages, and shapefiles.

ArcInfo applications

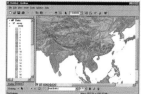

polymorphism

data components

geodatabase data access objects

data sources

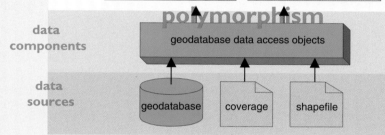

geodatabase | coverage | shapefile

extensible features

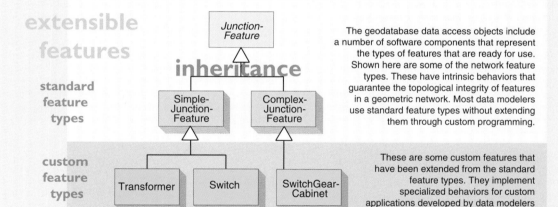

inheritance

Junction-Feature

Simple-Junction-Feature — Complex-Junction-Feature

standard feature types

The geodatabase data access objects include a number of software components that represent the types of features that are ready for use. Shown here are some of the network feature types. These have intrinsic behaviors that guarantee the topological integrity of features in a geometric network. Most data modelers use standard feature types without extending them through custom programming.

custom feature types

Transformer | Switch | SwitchGear-Cabinet

These are some custom features that have been extended from the standard feature types. They implement specialized behaviors for custom applications developed by data modelers and programmers.

data access

Data can be viewed in three ways.

The relational table view of data exposes the internal details of the physical storage as database tables.

The simple feature view presents data in the form of features without the structure of topology and relationships.

The object view of data encapsulates the internal details and presents a higher level of structure that is closer to the user's conceptual model of data.

relational table view of data

rules, domains

relational table

relationships

geometry column | attribute columns

simple feature view of data

geometric shapes with attributes

points

lines

polygons

object view of data

encapsulation

Dataset

Feature-Dataset | Table

ObjectClass | Relationship-Class

Feature-Class

ArcInfo accesses geographic data served through ArcSDE™, the Arc Spatial Database Engine. ArcSDE is the software technology that enables you to create geodatabases that range from small to very large sets of geographic data, and provides an open interface to the relational database of your choice.

HOW A GEODATABASE EXTENDS A DATABASE

These are some of the facets of a geodatabase that enhance relational database technology:

- A geodatabase can represent geographic data in four manifestations: discrete objects as vector features, continuous phenomena as rasters, surfaces as TINs, and references to places as locators and addresses.

- A geodatabase stores shapes of features and ArcInfo provides functions for performing spatial operations such as finding objects that are nearby, touching, or intersecting. A geodatabase has a framework for defining and managing the geographic coordinate system for a set of data.

- A geodatabase can model topologically integrated sets of features such as transportation or utility networks and subdivisions of land based on natural resources or land ownership.

- A geodatabase can define general and arbitrary relationships between objects and features.

- A geodatabase can enforce the integrity of attributes through domains and validation rules.

- A geodatabase can bind the natural behavior of features to the tables that store features.

- A geodatabase can present multiple versions so that many users can edit the same data.

PERSONAL AND MULTIUSER GEODATABASES

Geodatabases comes in two variants—personal and multiuser.

Personal geodatabase support is built into ArcInfo and is suitable for project-oriented GIS. A personal geodatabase is implemented as a Microsoft® Access database. When you install ArcInfo, Microsoft Jet is also installed; this provides the services for ArcInfo to create and update Access databases. You do not need to separately install Microsoft Access.

For large enterprises, you can deploy multiuser geodatabases with ArcSDE—the multiuser data access extension to ArcInfo. ArcSDE is installed on a data server that administers your organization's relational database. Through a TCP/IP network, ArcSDE serves geodatabases to the ArcInfo applications running on personal computers. ArcSDE can be run on Windows NT® or UNIX®.

ArcSDE allows remote access to geographic data and allows many users to view and edit the same geographic data. ArcSDE is centrally tuned and managed by your database administrator.

AN OPEN AND SCALABLE DATA SERVER

ArcInfo allows you to configure and deploy small to very large geodatabases. If you are working with moderately sized datasets, you can deploy personal geodatabases in ArcCatalog. This configuration yields good performance for datasets up to approximately 250,000 objects and supports one editor and several simultaneous viewers.

For more demanding datasets and to support many concurrent editors, you can deploy ArcSDE on the relational database best suited to your organization's needs.

These are some reasons to add ArcSDE to your ArcInfo installation:

- You have limitless flexibility in scaling databases.

- You can deploy the relational database of your choice.

- You can serve geographic data from UNIX or Windows NT.

- You can serve data to other applications such as MapObjects®, ArcIMS™ (Arc Internet Map Server), ArcView® GIS, and CAD client applications.

- You can centrally store and administer geodatabases.

- You can build Open GIS Consortium (OGC)-compliant applications.

- You can build Structured Query Language (SQL) applications to access the tables and rows in a geodatabase.

Open data framework

A geodatabase is an instance of a relational or object-relational database that has been enhanced by adding geographic data storage, referential integrity constraints, map display, feature-editing, and analysis functions.

geographically enhanced databases

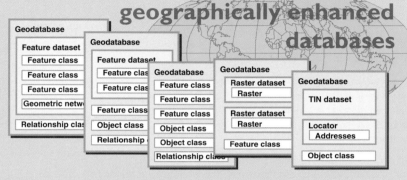

Geodatabases on any supported relational database operate identically.

Personal geodatabase

Personal geodatabase support is built into ArcInfo and provides access to local data.

ArcSDE

ArcSDE is a technology that uses the native data types and operators in a relational or object-relational database and extends them to provide the complete functionality of a geodatabase.

ArcSDE is the multiuser extension to ArcInfo.

relational databases

Microsoft Access	Oracle 8	SQL Server	Informix	DB2	Sybase

project GIS

A personal geodatabase is directed toward personal or small work-group use. It can handle small to moderately sized datasets.

Personal geodatabases are implemented on the Microsoft Jet engine, which stores data as Microsoft Access databases.

A personal geodatabase has all the functionality of a geodatabase served through ArcSDE, except for versioning.

enterprise GIS

A geodatabase served through ArcSDE can manage very large sets of geographic data and serve large numbers of viewers and editors. Geographic data is accessed from a data server on a network. This GIS data is centrally administered in large databases and integrates well with other corporate data. These databases require a system administrator for permissions, tuning, and optimization.

ArcSDE operates on any leading relational database. The ArcInfo developer can interact with a geodatabase through the geodatabase data access objects. A developer can access an ArcSDE geodatabase outside of ArcInfo through a C API (application programmer interface) or an SQL API.

To model work-flow processes, a geodatabase served through ArcSDE supports long transactions and version management. A versioned geodatabase allows many editors to work concurrently and includes a framework for resolving edit conflicts.

A developer can access data in a geodatabase at three basic levels:

- Through the geodatabase data access objects, a subset of ArcObjects™, the software components on which ArcMap and ArcCatalog are built.

- As simple nontopological features through the ArcSDE application programmer interface that complies with the OGC simple feature specification.

- As raw rows, columns, and tables through the native SQL interface of the relational database.

ACCESSING DATA THROUGH ARCOBJECTS

The richest level of accessing data is through the geodatabase data access objects. At this level, the full structure of a geodatabase is revealed: topology, relationships, integrity rules, and behavior, as well as raster, surface, and location representations.

You can programmatically access data through ArcObjects using Microsoft Visual Basic® for Applications (VBA) or with Visual C++® or other COM-compliant development environment.

The following is a simplified Unified Modeling Language (UML) diagram of a portion of the geodatabase data access objects. This is discussed in chapter 4, "The structure of geographic data."

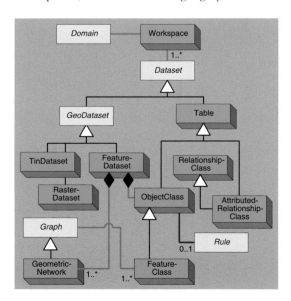

ACCESSING DATA AS SIMPLE FEATURES

For many spatial applications, it is sufficient and desirable to access geographic data in the form of simple nontopological features. This approach is especially suitable for building integrated applications for which geographic data is a vital component, but perhaps not the focus. Examples include facilities management and traffic analysis.

ArcSDE presents a simple feature API in C and Java™ that is compliant with the OGC simple features specification.

OGC is an organization of leading spatial data vendors, and its purpose is to develop standard software interfaces for the free exchange of spatial information among heterogeneous GISs.

Organizations that have geographic data in various formats on a network can build applications that integrate this data in the form of simple features.

ESRI is a leading contributor to the OGC technical specifications and is committed to the open exchange of geographic data.

ACCESSING DATA THROUGH SQL

A GIS is a rich repository of data about natural features or facilities such as transportation or utility networks. While this data is collected and managed as a geodatabase, external database applications can effectively access and share this data for nongeographic use.

Using the native SQL interfaces of your relational database, you can build applications to mine data from your geodatabases and use them for tasks such as managing inventory, processing work orders, or statistical analysis.

In this view, a geodatabase is a set of tables, rows, and columns. Through the SQL interfaces, you can see the internal database structure of a geodatabase, which includes metadata tables for objects such as networks. This structure is not directly visible in ArcInfo and is managed through the user interface of ArcCatalog. You can selectively update attributes of rows that represent features, but you should take care not to corrupt the structure of the geodatabase.

Accessing geodatabases

Developers can access a geodatabase through the geodatabase data access objects in ArcInfo, through APIs that expose simple features, or by the internal tables.

Geodatabase

Geodatabase
- Feature class
- Feature class
- Object class
- Relationship class
- Raster dataset
  - Raster

ArcInfo developer

through the geodatabase data access objects in ArcInfo and ArcSDE

GIS applications

ArcInfo is a general-purpose GIS application with advanced editing and map display, spatial analysis, and topological processing. Through ArcInfo, features in your geodatabase act with full object awareness as expressed with domains, validation rules, and custom code. The developer uses the geodatabase data access objects in Visual Basic, Visual C++, or other COM-compliant development environments.

Geometry

Point — Multipoint — *Curve*

Segment — Path — *Polycurve*
- Line
- EllipticArc
- CircularArc
- BézierCurve
- Ring
- Polyline
- Polygon

Spatial application developer

through ArcSDE API compliant with OGC simple features

spatial applications

Some applications process spatial queries from a large geodatabase and serve highly specialized functions. Examples are emergency response and business location. Geodatabases can be accessed as simple features though the ArcSDE simple feature API. This includes both C and Java APIs. The ArcSDE simple feature API is open and adheres to the OGC simple feature specification. This allows features in a geodatabase to be used outside of ArcGIS applications.

Database developer

through SQL interface in relational databases

database applications

Database applications sometimes need to extract data from a geodatabase, but not to display or spatially process that data. An example would be to pull or join utility pole attributes from a geodatabase to a relational database so that an inventory can be taken. The database programmer can interact with the tables in a geodatabase through the native SQL interfaces. The developer should refrain from modifying any geographic shapes or geodatabase system tables.

Personal geodatabase

ArcSDE

relational databases

Microsoft Access | Oracle 8 | SQL Server | Informix | DB2 | Sybase

Designing a geodatabase is fundamentally the same as designing any database. That is because a geodatabase is an instance of a relational database—one that contains a structure for representing geographic data.

The geodatabase extends, yet simplifies, the design process by presenting an object-oriented data structure that expresses the spatial and topological relationships of geographic features. Part of this structure is a special facility for representing sets of objects as integrated systems—such as stream and road networks or sets of land parcels. This structure on a set of features is called topology.

The geodatabase data model is the bridge between people's cognitive perception of the objects surrounding them in the world and how those objects are stored in relational databases.

GEODATABASE DESIGN

Traditional relational database design spans two basic steps—the articulation of a logical data model and the physical implementation of database models (or schemas).

The logical data model captures the user's view of data and the database model implements the data model within the framework of relational database technology.

Designing a logical data model

The key task in building a logical data model is to precisely define the set of objects of interest and to identify the relationships between them.

Some examples of objects you might consider are streets, parcels, owners, and buildings. Some examples of their relationships are "located at," "owned by," and "is part of."

Once you have an initial logical data model, you can validate it against the user's requirements for entering, updating, and accessing data and by testing it against the organization's practices and procedures (or business rules).

It is especially important to involve representatives from each prospective user group. A logical data model built for a subset of users is guaranteed to have deficiencies for overlooked users.

Building a logical data model is an iterative process and an art that is acquired through experience. There is no single "correct" model, but there are good models and bad models. It is difficult to determine precisely when a model is correct and complete, but an indication that you are coming close is when you can answer "yes" to these questions:

- Does the logical data model represent all data without duplication?

- Does the logical data model support an organization's business rules?

- Does the logical data model accommodate different views of data for distinct groups of users?

Representing logical data models

In the past, logical data models were often drawn in what are known as entity-relationship diagrams. A number of leading object-oriented modelers put forward various design methodologies and diagram notations.

These methodologies emphasized different aspects such as data flow or use-case scenarios, but a problem with entity-relationship diagrams is that their appearance varied with the design methodology.

More recently, most object-oriented modelers have adopted the Unified Modeling Language (UML), which is a standard notation for expressing object models and is endorsed by leading software and database companies.

It is important to note that UML is not a design methodology, but rather a diagrammatic notation. With UML, you can adopt the object-oriented design methodology of your choosing and express the model in a standard way.

This book uses UML for drawing that ArcInfo object model, called ArcObjects, and for drawing the custom object models you can create in a geodatabase.

Implementing a physical database model

A physical database model is built from the logical data model. Typically, a specialist in relational databases receives the logical data model from the data modeler and uses the database administration tools to define the database schema and create new databases ready for data transfer and entry.

The physical database design has some similarity to the logical data model, but there are differences. Classes of objects may be split or joined when implemented in tables. Rules and relationships can be expressed in several ways.

An important benefit of the geodatabase is that it is a physical implementation of data, but lets you structure your data in a fashion that is close to the logical data model.

Elements of the logical and database models

These are the basic elements of the logical data model and their corresponding database elements:

Logical elements	Database elements
Object	Row
Attribute	Column, Field
Class	Table

A logical data model is an abstraction of the objects that we encounter in a particular application. This abstraction is converted into database elements.

An object represents an entity such as a house, lake, or customer. An object is stored as a row.

An object has a set of attributes. Attributes characterize the qualities of an object, such as its name, a measure, a classification, or an identifier (or key) to another object. Attributes are stored in a database in columns (or fields).

A class is a set of similar objects. Each object in a class has the same set of attributes. A class is stored in a database as a table. The rows and columns in a table form a two-dimensional matrix.

Handling complex data

Relational databases enjoy their commercial dominance because they implement a simple, elegant, and well-understood theory. This simplicity is at once a strength and a weakness—it is conceptually straightforward to build relational databases, but difficult to model complex data.

Geographic databases contain complex data. The shapes of line and area features are structured sets of coordinates that cannot be well represented with standard atomic field types such as integer, real, and string. Further, features are gathered into systems that have explicit topological relationships, implicit spatial relationships, or general relationships.

The relational database is the foundation for a geodatabase. A key purpose of the geodatabase is to handle complex geographic data with a uniform data model independent of the relational database underneath.

Chapter 12, "Geodatabase design guide," returns to these topics in the context of designing and building geodatabases.

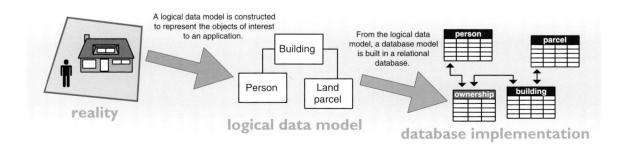

A logical data model is constructed to represent the objects of interest to an application.

From the logical data model, a database model is built in a relational database.

reality

logical data model

database implementation

GUIDELINES FOR GEODATABASE DESIGN

The structure of a geodatabase—feature datasets, feature classes, topological groupings, relationships, and other elements—lets you design geographic databases that are close to their logical data models. For a data modeler, this is the essential reason for the introduction of geodatabases into ArcInfo 8.

These are the basic steps in designing a geodatabase:

1 *Model the user's view of data.* Perform interviews with users, understand an organization's structure, and analyze the business requirements.

2 *Define objects and relationships.* Build the logical data model with the set of objects, knowing how they are related to one another.

3 *Select geographic representation.* Determine whether vector, raster, surface, or location representation is best for the data of interest.

4 *Match to geodatabase elements.* Fit the objects in the logical data model into the elements of a geodatabase.

5 *Organize geodatabase structure.* Build the structure of a geodatabase with consideration of thematic groupings, topological associations, and department responsibility of data.

This topic is discussed in greater detail in chapter 12, "Geodatabase design guide."

Steps to building a geodatabase

Model the user's view of data.

Identify organizational functions.
Determine data needed to support functions.
Organize data into logical groupings.

Define objects and relationships.

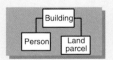

Identify and describe objects.
Specify relationships between objects.
Document model in diagram.

Select geographic representation.

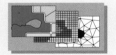

Represent discrete features with points, lines, areas.
Characterize continuous phenomena with rasters.
Model surfaces with TINs or rasters.

Match to geodatabase elements.

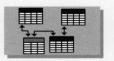

Determine geometry type of discrete features.
Specify relationships between features.
Implement attribute types for objects.

Organize geodatabase structure.

Organize systems of features.
Define topological associations.
Assign coordinate systems.
Define relationships and rules.

You can approach ArcInfo in two ways: as a user of applications such as ArcMap and ArcCatalog, or as a software developer building custom applications.

Data modelers straddle these two worlds—you use the applications for most of your work in creating geodatabases, but you will sometimes write software code, especially if you are trying to create rich data models that support powerful applications.

One aim of this book is to present the important data-modeling concepts both as they are applied in the ArcInfo applications and in the ArcInfo software components, called ArcObjects.

A pattern throughout this book is to first present the concepts for a topic as you experience it through the ArcInfo application. Next, that topic is summarized with an annotated diagram of the relevant section of the ArcInfo object model diagram.

For example, the topic of the structure of geodatabases, feature datasets, and feature classes is first discussed from the user's perspective within ArcCatalog. Next, the programmer's perspective is summarized in a diagram of part of the geodatabase data access objects.

These two views have similarities, but also subtle differences. A user interface sometimes hides details about software components that are important to the programmer. One goal of this book is to give you the insight to bridge the user and developer perspectives.

Reading the class diagrams

This is the key for the object model diagrams you will find throughout this book:

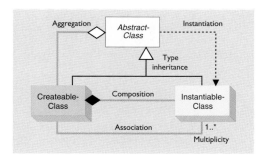

This notation is based on the UML notation, an industry diagramming standard for object-oriented analysis and design.

The object model diagrams are an important supplement to the information you receive in object browsers. The development environment, Visual Basic or other, lists all of the many classes and members, but does not hint at the structure of those classes. These diagrams complete your understanding of the ArcInfo components.

This book uses UML to document the ArcInfo software components, ArcObjects, and to illustrate custom data models that you can build.

Classes and objects

There are three types of classes shown in the UML diagrams—abstract classes, createable classes, and instantiable classes.

An *abstract class* cannot be used to create new objects, but it is a specification for subclasses. An example is that a "line" could be an abstract class for "primary line" and "secondary line" classes.

A *createable class* represents objects that you can directly create using the object declaration syntax in your development environment. In Visual Basic, this is written with the *Dim As New <object>* or *CreateObject(<object>)* syntax.

An *instantiable class* cannot directly create new objects, but objects of this class can be created as a property of another class or created by functions from another class.

In the Visual Basic object browser, you can inspect all of the ArcInfo createable and instantiable classes, but not the abstract classes.

Relationships

Among abstract classes, createable classes, and instantiable classes, there are several types of class relationships possible.

Associations represent relationships between classes. They have defined multiplicities at both ends.

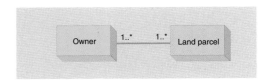

In this diagram, an owner can own one or many land parcels and a land parcel can be owned by one or many owners.

A *Multiplicity* is a constraint on the number of objects that can be associated with another object. This is the notation for multiplicities:

1—One and only one. Showing this multiplicity is optional; if none is shown, "1" is implied.

0..1—Zero or one

M..N—From M to N (positive integers)

* or 0..*—From zero to any positive integer

1..*—From one to any positive integer

Type inheritance defines specialized classes that share properties and methods with the superclass and have additional properties and methods.

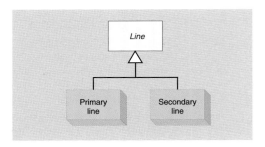

This diagram shows that a primary line (createable class) and secondary line (createable class) are types of a line (abstract class).

Instantiation specifies that one object from one class has a method with which it creates an object from another class.

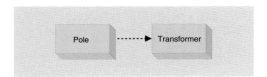

A pole object might have a method to create a transformer object.

Aggregation is an asymmetric association in which an object from one class is considered to be a "whole" and objects from the other class are considered "parts."

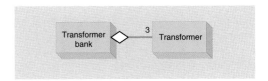

A transformer bank has exactly three transformers. In this design, transformers can be associated with a transformer bank, but may also exist after the transformer bank is removed.

Composition is a stronger form of aggregation in which objects from the "whole" class control the lifetime of objects from the "part" class.

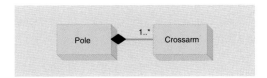

A pole contains one or many crossarms. In this design, a crossarm cannot be recycled when the pole is removed. The pole object controls the lifetime of the crossarm object.

Expressing models with diagram notation

If you are unaccustomed to this type of diagram notation, practice reading the examples above and conceive of your own examples. Before long, you will read these diagrams with ease. You will find that it is worth your effort to understand this notation. It describes object models in a concise and expressive way and will facilitate your conceptual understanding of the ArcInfo software components.

Understanding this notation is also critical if you create custom features by extending the geodatabase data access objects. With ArcCatalog, you can launch a computer-aided software engineering (CASE) environment to create custom data models with a visual user interface. This interface is based on manipulating graphical symbols from the UML notation.

A geographic information system is at its core a database management system enhanced to store, index, and display geographic data.

ArcInfo 8 is a significant release of new GIS technology that exploits several important technology trends just as they have become ready for commercial implementation. These trends collectively realize the vision of GIS as a geographically enabled database.

The timing of ArcInfo 8 is fortuitous as it occurs during the convergence of several critical developments in software and database technology. The following are the principal trends that shape the technological framework of ArcInfo 8.

Spatial data and databases

When the coverage data model was first implemented, practical considerations led to the spatial component of geographic data being contained in binary files with unique identifiers to rows in relational database tables that stored feature attributes.

With performance and functional advances in database technology, it is now possible and advantageous to store all spatial data directly within the same database tables as attribute data.

The gain from storing spatial data directly within commercial databases is improved data administration, the utilization of data access and management services, and closer integration with the other databases that an organization manages.

Moreover, ArcInfo users can select from any of the industry-leading relational databases to host their geographic databases.

User interface

Applications developed for Microsoft Windows® have set a new standard for ease of use and consistency. Users have become accustomed to expected behaviors for mouse interaction, menus, dialog boxes, and the like. These user interface standards have made powerful applications accessible and usable by people who are not computer experts.

ArcInfo 8 thoroughly implements the Windows standards for user interface and stands as a new milestone in making GIS software easier to use.

Software component architecture

Modern software is built on software component architectures, examples of which are Microsoft Component Object Model (COM), the Common Object Request Broker Architecture (CORBA), and Java Remote Method Invocation (RMI).

The idea behind components is to divide software functionality into discrete, independent pieces that can be developed, tested, and combined into programs. By their design, components can be used to build any number of applications without modification. This is a high level of software reuse.

The benefit of software component architectures is better software quality, better performance, and the ability to update software versions without affecting other installed software.

ArcInfo 8 is built on the Microsoft COM architecture because it is the most robust and reliable component framework for desktop applications.

Programming environment

Visual programming environments such as Visual Basic have become the norm for application development.

The benefits of using these languages are the large pool of experienced programmers and the richness of these environments. It is no longer necessary or desirable to use proprietary macro languages.

ArcInfo 8 uses Visual Basic for Applications (VBA) as its embedded macro language for customizing its applications, ArcMap and ArcCatalog. Other COM-compliant languages such as Visual C++ can be used to extend the geodatabase data model.

Trends in summary

The common themes of these technology trends are open standards and interoperability.

The benefit of implementing these trends is to take advantage of technology from other industry segments, which lets ESRI concentrate its research and development on core GIS functionality.

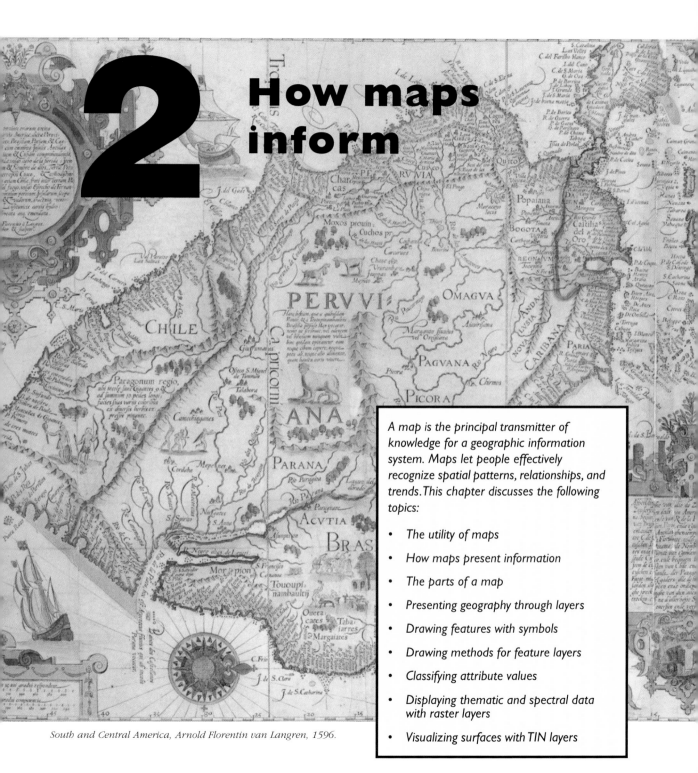

2 How maps inform

A map is the principal transmitter of knowledge for a geographic information system. Maps let people effectively recognize spatial patterns, relationships, and trends. This chapter discusses the following topics:

- The utility of maps

- How maps present information

- The parts of a map

- Presenting geography through layers

- Drawing features with symbols

- Drawing methods for feature layers

- Classifying attribute values

- Displaying thematic and spectral data with raster layers

- Visualizing surfaces with TIN layers

South and Central America, Arnold Florentin van Langren, 1596.

People have used maps throughout history. Until recently, maps were exclusively printed documents. Drawn on flat sheets of paper or parchment, maps depicted objects in the real world—paths, settlements, and natural features.

The practice of cartography evolved to support diverse and inventive ways to characterize the many qualities of the real world. Techniques were developed to depict classifications of features, identifying labels, the shape of the earth's surface, and the flow of resources or goods.

Many of these practices are manifest in our modern maps, such as the use of double-line symbols for roads, text label placement, and the application of the color blue for bodies of water.

With the widespread adoption of computers and the development of GIS technology, maps are now the printed documents with which we are familiar, as well as interactive visual displays on computers.

GIS systems have enhanced the way people interact with maps. You can easily define the manner in which information is presented and can also select locations or objects to initiate a query or analysis.

WHAT MAPS DO

Maps are uniquely capable for sharing knowledge about our world in many ways.

Maps identify what is at a location. You can point to a location on a map and learn the name of the place or object and any other descriptive attributes.

Maps can locate where you are. If your map has real-time input from the Global Positioning System (GPS), you can see where you are, how fast you are traveling, and the direction you are headed.

Maps let you identify distributions, relationships, and trends not otherwise discernible. A demographer can compare maps of urban areas compiled in the past with present-day maps to guide public policy. An epidemiologist can correlate the locations of rare disease outbreaks with environmental factors to find possible causes.

Maps can integrate data from diverse sources into a common geographic reference. A municipal government can merge street maps with maps from utilities to coordinate construction. An agricultural scientist can couple images from weather satellites with maps of farms and crops to boost productivity.

Maps let you combine and overlay data to solve spatial problems. A state or provincial government can combine many layers of data to find suitable locations for a waste disposal site.

Maps can find the best path between one place and another. A package delivery firm can find the most efficient route for trucks. A public transportation planner can create optimized bus routes.

Maps can model future events. A utility company can simulate the impact of a new subdivision and determine the necessary system upgrades. A regional planner can model serious accidents such as a toxic spill and develop evacuation scenarios.

WHAT MAPS ARE

GIS technology has broadened our view of a map. Instead of a static entity, a map is now a dynamic presentation of geographic data.

A map is the graphical presentation of geographic data. To be effective, a map must be visually compelling. Principles of graphic design—layout, proportion, balance, symbology, and typography—apply to maps as well as to other types of illustration.

A map is the interface between geographic data and our perception. Maps utilize people's inherent cognitive abilities to identify spatial patterns and provide visual cues about the qualities of geographic objects and locations.

A map is an abstraction of geographic data. A map is a view of geography for a particular class of user. A map filters information for intended use—only information for the intended purpose is displayed. A map simplifies data—some of the complexity and internal structure of data is hidden. A map adds descriptive content to data—labels reveal names, categories, types, and other information.

The goal of a data modeler is to design a data structure that supports the creation of informative and aesthetic maps. Understanding how maps inform is the prerequisite to building a data model.

When you read a map, you observe facts about the shape and position of geographic features, the attribute information associated with geographic features, and the spatial relationships among features.

HOW MAPS EXPRESS GEOGRAPHIC INFORMATION

Geographic features are located at or near the surface of the earth. They can occur naturally (rivers, vegetation, and peaks), can be constructions (roads, pipelines, and buildings), and can be subdivisions of land (counties, land parcels, and political divisions).

Three primary ways of presenting a geographic area on a map are as a set of discrete features, as an image or sampled grid, and as a surface.

DISPLAYING DISCRETE FEATURES

Many geographic features have distinct shapes that can be portrayed by points, lines, and polygons.

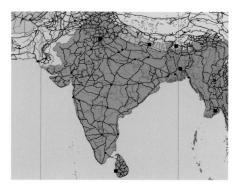

Points represent geographic features too small to be depicted as lines or areas, such as well locations, telephone poles, and buildings. Points can also represent locations that have no area, such as mountain peaks.

Lines represent geographic features too narrow to be depicted as areas, such as streets and streams, or slices through a surface, such as elevation contours.

Polygons are closed figures that represent the shape and location of homogeneous features, such as states, counties, parcels, soil types, or land-use zones.

DISPLAYING IMAGES AND SAMPLED GRIDS

Much of the information we collect about the earth is in the form of *aerial photographs* or *satellite images*. These images often form a backdrop to other map data.

Similar in format to images are *sampled data grids,* which represent a continuous phenomenon such as temperature, rainfall, or elevation.

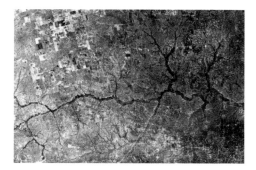

Images and sampled data grids are called *rasters*. A raster is comprised of a two-dimensional matrix of *cells,* which have attributes that represent qualities such as color, spectral reflectance, or rainfall.

DISPLAYING SURFACES

The shape of the earth's surface is continuous. Some aspects of a surface can be drawn as features, such as ridges, peaks, and streams. Lines of equal elevation can be drawn as contour lines.

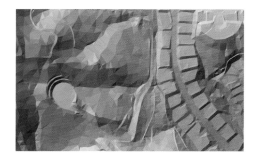

To portray the shape of the earth, you can create a surface display that uses a range of colors to characterize sun illumination, elevation, slope, and aspect. Most often, the vertical values represent an elevation, but other attributes such as population density can define a surface as well.

HOW MAPS PORTRAY ATTRIBUTES

The features on a map have any number of associated attribute values. These attributes reside within the database table for a set of features or can be accessed through links to other databases.

The most common types of attributes are these:

- A *descriptive string* gives a feature its name or characterizes a category, condition, or type.
- A *coded value* represents a type of feature. It can be a numeric value or an abbreviated string.
- A *discrete numeric value* represents something that is counted, such as the lanes on a road.
- A *real numeric value* represents continuous data that is measured or calculated, such as distance, area, or flow.
- An *object identifier* is rarely displayed, but it is the key to access attributes in external databases.

There are a variety of techniques for illustrating descriptive information on a map.

Depicting type attributes

Coded values are used to draw symbols that depict a type of object. Points are drawn with recognizable symbols for schools, mines, and ports. Lines are drawn with distinct pen patterns that represent continuous or intermittent streams. Areas are drawn with fill patterns that portray any classification.

Illustrating measured attributes

Numeric values can be drawn on a map by varying the size of symbols. These values can be integers or real numbers and can be grouped into classifications.

Drawing classified attributes

Coded values or numeric values can be presented on a map by using colors. A color can represent the features that share a common value. A color can represent a numeric value within a range by a blend from one color to another or a gradation in hue, brightness, or saturation.

Labeling descriptive attributes

Descriptive strings can be drawn next to, along, or inside the features they describe.

HOW MAPS EXHIBIT SPATIAL RELATIONSHIPS

When you look at a map, your mind discerns spatial patterns. Many maps are built for purposes such as identifying business locations, optimizing routes, and understanding habitats.

Maps visually reveal these spatial relationships:

- Which features connect to others
- Which features are adjacent to others
- Which features are contained within an area
- Which features intersect
- Which features are near others
- The difference in elevation of features
- The relative position among features

Maps in a GIS also support spatial queries that create lists and selections.

ArcInfo and its mapping application, ArcMap, present a model of a digital map that conforms to our experience with traditional maps.

You can print this digital map on a large-format printer to high cartographic standards. You can interact with this digital map on a computer and modify the thematic display, query features, perform analysis, and edit features. The digital map is stored as a file with an extension of .mxd and is called a *map document,* or simply, a *map.*

THE MAP AND ITS ELEMENTS

A map document contains cartographic elements with which you are already familiar—north arrows, scale bars, neatlines, titles, insets, and legends. The main elements of a map are organized this way:

* A map has one or more *data frames* that present geographic data.

* Each data frame has one or several *map surrounds* that display a cartographic context.

* The page layout of a map has a number of *map elements* that finish the map.

The map's container of geography

A *data frame* contains the geographic data on your map. A map can have one or several data frames.

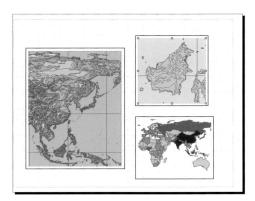

A data frame has one or many *layers* that are stacked on top of each other and span the same geographic extent. A data frame occupies an extent on the page layout and spans a geographic extent. The ratio between a data frame's geographic extent and its layout extent is the *map scale.*

A data frame has a *coordinate system* that describes how that part of the world is projected. This coordinate system may be the same as or different than the coordinate system of the layers.

The cartographic qualities of a data frame

A data frame can be associated with *map surrounds* that present the cartographic context such as scale.

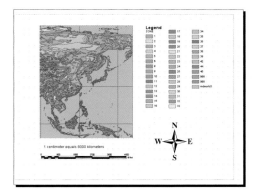

Map surrounds are dynamically linked to a data frame. When the drawing method is changed, the legend is updated. When the map scale is changed, scale text is updated and the scale bar is resized. When the map is tilted, the north arrow is rotated.

The finishing graphics of a map

You can add *map elements* to complete your map.

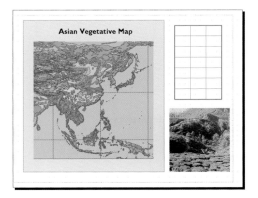

Map elements include markers, lines, polygons, rectangles, text, and pictures. A picture can be a Windows metafile or bitmap. Map elements have no explicit association with the data frame.

A *layer* is the basic unit of geographic presentation on a map. It shows a set of related geographic data drawn to a cartographer's specifications. Some examples of layers you might create are streams, political boundaries, survey points, and roads.

LAYERS ABSTRACT GEOGRAPHIC DATA

A layer is a reference to a set of geographic data, but it does not contain geographic data. There are several advantages to this approach:

- You can create distinct layers on the same geographic data that visualize different attributes or employ different drawing methods.

- You can edit geographic data, and map layers are updated the next time you display the map.

- Layers are shared across an organization without duplicate geographic data. A layer can reference data at any location accessible on a network.

A layer is stored as a part of a map document or as a separate file on your computer disk with an extension of .lyr. You can think of a layer as a cartographic view of geographic data. A layer lets you assign drawing methods, set scale thresholds, and apply selections to the display.

Drawing many views of geographic data

A layer lets you assign any type of drawing method to a geographic dataset.

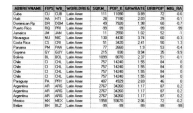

A geographic dataset of the world's countries might have a number of attributes such as population, life expectancy, growth rate, and water quality.

However, geographic datasets do not contain the instructions for drawing the data. You specify the methods for drawing data when you create a layer.

You can create multiple layers for the same dataset. Each of these layers can depict a separate attribute.

These maps show life expectancy, water quality, and population growth in South America.

Drawing selections of features

Some maps show subsets of features in a dataset. When you create a layer, you can select features interactively on the map or specify an attribute query using Structured Query Language (SQL) syntax.

The first map shows all the countries in Europe; the second shows those countries participating in currency unification.

With selections in a layer, you can draw only the features of interest without having to delete features.

Controlling the map scale of layers

You can draw a map to any map scale, but certain layers are best drawn within a prescribed scale range. You can set a scale threshold for a layer and replace one layer with another at a specified scale.

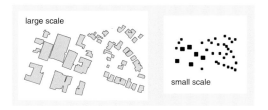

The first map shows a layer with buildings drawn with fill symbols. The second map shows a layer with the same geographic dataset, but drawn with marker symbols.

TYPES OF LAYERS

Recall that a geographic area can be presented on a map as a set of discrete features, as images or grids, or as surfaces. Below are some of the types of layers you can add to a map.

Most layer types are associated with geographic datasets within geodatabases. Subsequent chapters of this book contain further information on these data objects.

Mapping discrete features

Many geographic objects have a distinct shape.

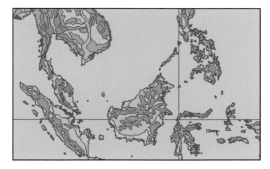

A *feature layer* uses a drawing method to present descriptive information about a feature class. A *feature class* is a homogeneous collection of point, line, or polygon features.

Mapping images and sampled grids

Much of the geographic data that is collected is in the form of satellite imagery, photographs, or grids.

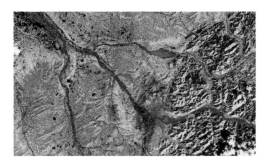

A *raster layer* uses a drawing method to present spectral or descriptive information about a raster. A *raster* is a matrix of cells with attribute values.

Mapping surfaces

Surfaces represent the shape of the earth.

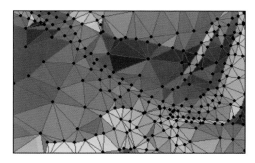

A *TIN layer* uses a drawing method to show the z value of a triangulated irregular network (TIN). A TIN is composed of adjacent triangles that share nodes and edges.

Maps present descriptive information about geographic features using symbols and labels.

Here are some common ways that maps present descriptive information about the geographic features they represent:

- Roads are drawn with various widths, patterns, and colors to represent different road classes or other attributes.

- Streams and water bodies are typically drawn in blue to indicate water.

- Special symbols denote specific features, such as railways and airports.

- Streets are labeled with names.

- Buildings can be labeled with their name or function.

Four basic types of symbols are used to present descriptive information about features: marker symbols, line symbols, fill symbols, and text symbols.

Drawing points with marker symbols

You can choose from several types of marker symbols to represent point features on a map.

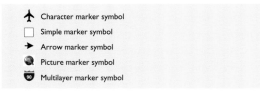

A *character marker symbol* is based on a single character (or glyph) in a TrueType® font. These symbols are drawn with one color.

A *simple marker symbol* is a predefined simple stroked symbol such as a square or circle optimized for rapid screen display.

An *arrow marker symbol* is based on a single character in a predefined TrueType font for the purpose of drawing arrows (or line decorations) at the ends of cartographic lines.

A *picture marker symbol* is a bitmap or enhanced metafile. A bitmap is a standard Windows® raster

image with a file extension of .bmp. An enhanced metafile is a standard Windows vector drawing with a file extension of .emf. Enhanced metafiles can have many colors and, because they are based on vector graphics, can be drawn at different sizes without visual degradation.

A *multilayer marker symbol* is a composite symbol that combines any of the other types of marker symbols. This is ideal for complex symbols that are a combination of shapes and text, such as highway shield symbols. A simple marker symbol can be used as an outline for a multilayer marker symbol.

Drawing lines with line symbols

Linear features on a map can be drawn with one of the following line symbols:

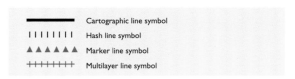

A *cartographic line symbol* is a general-purpose line symbol with display properties of width, color, parallel offset distance, dash pattern (or template), arrow heads (or line decoration), cap, and join. Cap specifies whether the ends of line symbols are drawn squared, butted, or rounded. Join specifies whether corners of lines are square, rounded, or beveled.

A *hash line symbol* has short segments that are perpendicular or at any specified angle to the path of a line. Hash line symbols are usually combined with cartographic line symbols within a multilayer line symbol; the customary symbol for railroad tracks is an example of this.

A *marker line symbol* contains marker symbols in a pattern defined by a template. Any type of marker can be placed within a marker line symbol.

A *multilayer line symbol* is a composite symbol that combines any of the other types of line symbols. The example of a railroad track symbol is achieved by combining a cartographic line symbol with a hash line symbol in a multilayer line symbol.

Drawing areas with fill symbols

You can draw areal features with one of these fill symbols:

A *simple fill symbol* has display properties of color, outline style (null, solid, dashed, and others), and outline width. A simple fill symbol can also contain a number of predefined line fill patterns such as horizontal hatch or crosshatch. Simple fill symbols can be hollow; you can draw areal features by outline only.

A *line fill symbol* has the properties of a simple fill symbol, but you can specify a richer type of line fill pattern that can incorporate any line symbol at any angle and separation.

A *marker fill symbol* is drawn either as a grid of marker symbols that can be arbitrarily spaced and rotated, or as a random distribution of marker symbols with a specified average horizontal and vertical separation.

A *gradient fill symbol* is drawn as a blend of two colors, transitioning from one to another. There are four types of gradient fill:

- A linear gradient blends colors in one direction, from top to bottom, left to right, or at any angle.

- A radial gradient blends colors in a circular pattern from the center point outward to the outer part of the area.

- A rectangular gradient blends colors from the center outward in a rectangular pattern.

- A buffered gradient blends colors inward from the perimeter of an area. A percentage value limits how far the gradient progresses inward from the perimeter. This is ideal for the cartographic convention of drawing ocean shorelines.

A *picture fill symbol* is comprised of bitmaps or enhanced metafiles. The pictures are drawn contiguously or with a fixed spacing.

A *multilayer fill symbol* is a composite symbol that combines any of the other types of fill symbols.

A *feature layer* is a reference to a feature class and has an associated *drawing method* (or *renderer*). You can choose any string or numeric attribute of your feature layer and visualize it in a variety of ways.

You will find that the type of attribute you are interested in visualizing guides your selection of a drawing method. Numeric data might be best presented with symbols that change size or color according to the attribute value. Attributes that describe a type of feature might be best drawn with symbols that match each unique value.

The following sections outline the drawing methods available for feature layers.

DRAWING FEATURES

The simplest way to draw a feature layer is to draw all the features with the same symbol.

With this drawing method, all features are drawn with a symbol that follows a cartographic convention. Well heads could be drawn with a square marker, streams could be drawn with blue lines, and buildings could be drawn with simple yellow fill symbols with a black outline.

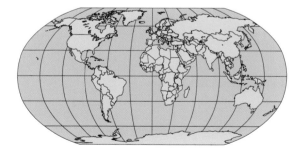

This map draws all countries with the same fill symbol.

This drawing method is suitable for feature layers that represent a fairly homogeneous set of geographic features. It is also used for a simple display of feature layers that are behind other layers of greater interest.

This method is also best for drawing features simply so that spatial distribution patterns can be visually recognized. If your map contains point features for potential customers, this drawing method can help you discern spatial clusters for geographically targeted marketing.

This drawing method is called the *simple renderer* in the ArcInfo object model.

DRAWING CATEGORIES OF FEATURES

The feature's attribute of interest can be drawn by creating categories. The following are the techniques used to symbolize by category.

Drawing categories by unique field values

An attribute in a feature layer sometimes represents an important subdivision of the feature type. This attribute can describe a category of a feature, such as a land-use type or type of road. It can also characterize a relation between the feature and a larger entity, such as a province or state and the country to which the feature belongs.

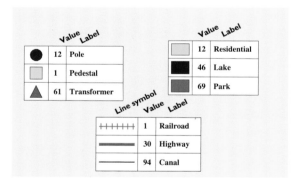

This drawing method lets you assign a unique symbol to each unique value of the attribute. An electric device layer can contain a type attribute representing poles, pedestals, and transformers. A transportation layer can have a type attribute for railroads, highways, and canals. A land-use layer can have a classification attribute designating residential, lake, or park status.

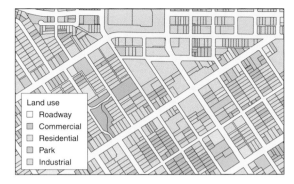

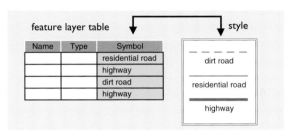

This map shows areas drawn with a unique symbol for each type of vegetation.

With this type of drawing method, you can surmise how different types of features are located with respect to each other and their relative frequency and distribution.

This drawing method is called the *unique value renderer* in the ArcInfo object model.

Drawing categories by unique combined field values

You can elect to draw categories of the unique combinations of up to three field values. In this planning map, you can see combined values of land use and historic district zones.

Use this drawing method with care. It is not difficult for the number of unique combined values to become too large to visually discern the classifications you want to differentiate.

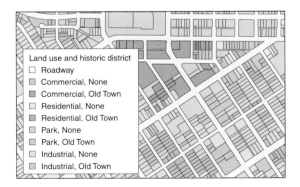

This map shows unique combinations of field values.

Drawing categories by symbol names in a field

You can draw features by using symbol names in a field. This field contains symbol names as text values such as "Primary road," "Industrial zone," or "Survey marker." This field can have any name.

This drawing method is the easiest way to ensure that symbols are drawn exactly the same way in different maps throughout an organization.

Another advantage of this drawing method is that an organization can implement parallel styles to produce distinct sets of maps. For example, you can create a set of styles with symbols intended for different map scales. At one map scale, a road can be drawn as a simple line, and at another scale, that road can be drawn with double lines. This drawing method makes it easy to switch how symbols are drawn for different map products. Features drawn with this method appear the same as features drawn by unique values.

DRAWING QUANTITIES OF FEATURES

Numeric fields can store values that are numerically ordered and can represent counted or continuous values. The following are the drawing methods to visualize the quantities of features.

Drawing quantities by color

An effective way to display a numeric attribute is to present an attribute with a set of graduated colors. This attribute, called a *value field*, can be *normalized* by another field. This means one value is divided by another.

The value field is subdivided into a set of classes. You have several options for *classification*, and this is discussed in some detail in the next chapter topic. Classification is a statistical process for subdividing a collection of values.

The graduated colors are set by selecting a *color ramp,* which is a transition from one color to another. A color ramp can have multiple parts—a beginning color can blend to an intermediate color and then to a final color.

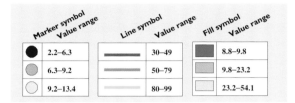

This drawing method is an effective presentation for numeric data that represents a continuous attribute such as elevation, temperature, or amount of a resource. It is especially effective for drawing areas.

Color ramps are used to apply cartographic practices. For example, bathymetric maps are drawn with light blue for shallow waters to dark blue for deep waters.

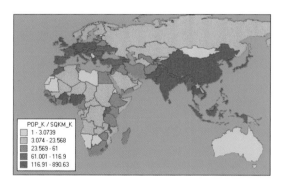

This map shows population density by drawing a population attribute normalized by an area attribute.

This drawing method is called the *class breaks renderer* in the ArcInfo object model.

Drawing quantities by graduated symbols

Another effective way to visually present a numeric attribute is to vary the size of a symbol. Again, you specify a value field, an optional normalization field, and a classification. The range of values is subdivided into the number of classes set in the classification.

Instead of a color ramp, you select a base symbol and a range of symbol sizes.

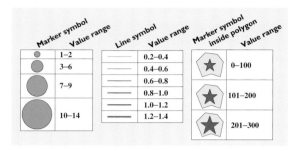

This drawing method is suitable for numeric data that represents a rank or progression of values. Some care should be taken in selecting the range of symbol sizes so that features do not overlap excessively in dense areas.

This method draws the larger symbols first and the smaller symbols afterward. This is so that features with large values do not obscure features with smaller values. You can choose contrasting colors for the outline and body of the symbol to make overlapping features stand out.

An interesting behavior of this drawing method is that value ranges for areas are drawn with marker symbols instead of fill symbols. That is because the size of an area is predetermined by its shape, and drawing a marker symbol at its centroid point is an alternative to drawing a graduated symbol.

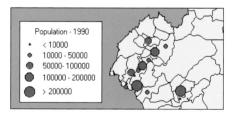

This map draws cities with marker symbols graduated by population value against a layer with administrative areas.

This drawing method is called the *graduated symbol renderer* in the ArcInfo object model.

Drawing quantities by proportional symbols

This drawing method is similar to drawing features with graduated symbols, except that there is no classification of values and the size of each symbol varies in exact proportion to the attribute value.

You specify a value field, an optional normalization field, and the units for display. You can also specify whether the attribute value varies the symbol's size by area or radius.

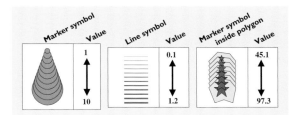

This drawing method is suitable when you want to draw a continuous gradation of symbol sizes.

As with the graduated symbol drawing method, symbols can overlap each other. Therefore, larger symbols are drawn first, followed by smaller symbols, and the values for area features are drawn with marker symbols.

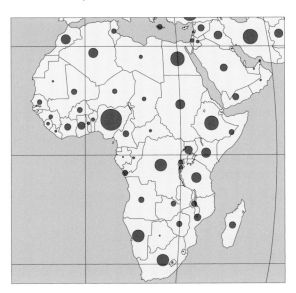

This map shows countries that are sized in exact proportion to their population.

This drawing method is called the *proportional symbol renderer* in the ArcInfo object model.

DRAWING MULTIPLE ATTRIBUTES

On occasion, you will want to symbolize features by multiple attribute values.

This drawing method lets you effectively use two renderers at once on a feature layer. This method is called the *bi-unique value renderer* in the ArcInfo object model.

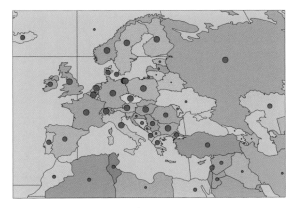

This map shows the countries of Europe drawn with two distinct attributes symbolized by color coding a unique value on the polygons and a quantitative value on the point symbols.

Two of the drawing methods for feature layers—drawing features with graduated colors and drawing features with graduated symbols—are based on a classification of attribute values. A drawing method for raster layers—drawing cells by graduated colors—also uses a classification of attribute values.

CLASSIFICATION METHODS

A *classification method* is applied toward a group of attribute values and subdivides them according to the desired criteria. These can be equal subdivisions of the range of attribute values, equal counts of features within each class, or another criterion.

Each subdivision of the group of attribute values is known as a *class*. Each class has a lower and upper numeric range limit. For each of these classes calculated by any of the classification methods, you can override a class range limit and set another.

Classifying values by natural groupings

The *natural breaks classification* uses a statistical formula to determine natural clusters of attribute values. The function of the formula, known as Jenk's method, is to minimize the variance within a class and to maximize the variance between classes.

The graduated symbol and graduated color drawing methods apply this classification method by default.

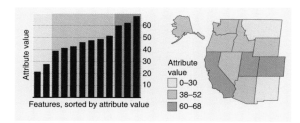

The natural breaks classification is well suited to uneven distributions of attributes. Distinct natural groupings of attributes can be isolated and highlighted.

Classifying values by defined intervals

The *defined interval classification* divides a set of attribute values into classes that are divided by precise numeric increments, such as 10, 100, or 500.

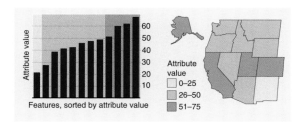

This classification works well for values that people are accustomed to seeing in rounded numbers, such as age distribution, income level, or elevation ranges. The disadvantage is that some of the classes, particularly the first and last, may contain a disproportionate number of feature values.

Classifying values by equal intervals

The *equal interval classification* takes the range of values and subdivides them into ranges of equal value intervals.

A value range from 21 to 69 with three classes would be subdivided into range spans of 16 units, 21–36, 37–52, and 53–69.

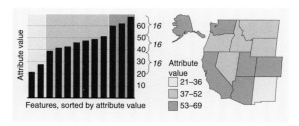

This classification emphasizes how feature values fall within uniform ranges of attribute values. In practice, it is similar to defined intervals, but has the advantage that the lowest and highest class span the same range of values as the rest of the classes.

An example of the application of this classification is a map that depicts homes for sale divided into equal ranges of purchase costs.

Classifying features by quantiles

The *quantile classification* creates classes with equal numbers of features. If a feature layer has 12 features, three classes would each represent four features.

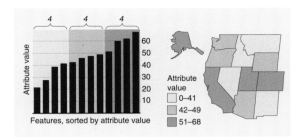

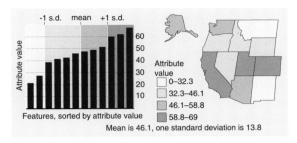

Mean is 46.1, one standard deviation is 13.8

This classification is particularly effective for ranked values. A company can measure sales performance of business locations and draw the respective businesses in their rank of sales performance. This classification yields visually attractive maps because all of the classes drawn have the same number of features.

However, this classification might obscure the natural distribution of attribute values; clusters of attribute values may be split or combined with other values. This classification is best applied to attribute values that have a generally linear distribution.

If you have an even number of classes, the value delimiting the middle classes is the same as the *median* of a statistical sampling.

Classifying features by standard deviations

The *standard deviation classification* creates an even number of classes that represent whole or fractional deviations from a mean value.

First, the *mean* (or *average*) of all the attribute values is calculated. Then a statistical formula calculates a standard deviation.

You can specify the number of classes and whether they span one whole standard deviation, one-half standard deviation, one-third standard deviation, or one-quarter standard deviation. The classes at the low and high end extend to the minimum and maximum values.

This classification is intended for generally symmetric distributions of attribute values that have a broad peak near the mean with the density of values diminishing away from the peak.

An example of a suitable map for this classification could be population density or accident rates. You would expect these values to have their greatest data density near a mean value and that values that vary significantly are scarce. The classic shape of this type of distribution is the bell curve.

Normalizing attribute values

Sometimes, a classification is best applied not to a single attribute, but to one attribute normalized by another. Normalization is simply dividing one attribute value by another.

An example of a normalized attribute would be accident rate. The data might contain accumulated values for sections of highways, but for accident data to be meaningful, the number of accidents should be normalized (or divided) by the length of each highway segment.

Excluding attribute values

Some data contains erroneous or null values, or you might want to examine only a certain range of values.

Erroneous data might be flagged as values outside a reasonable range. For example, percentage values might always be expected to be from zero to 100. Smaller and larger values would be excluded.

Attribute values that represent a ranking of features may have a coded value such as –99 that represents a null or unknown value. This value should be excluded.

Much of the most readily available geographic data is in the form of rasters. A *raster* is a regularly spaced matrix of cells that may have associated attribute information.

A *cell* can represent either continuous data such as elevation and pollution concentration, or spatially discrete data such as land use or vegetation type.

A raster can have a *single band* or *multiple bands*. A band is like a layer that represents different values for each cell. A common example of this value is light reflectance at a part of the spectrum.

A *raster layer* is a reference to a raster with a specified drawing method. The same raster can be drawn with several raster layers, each with a drawing method to emphasize a particular attribute or classification.

Chapter 9, "Cell-based modeling with rasters," contains more information on the raster data structure, its advantages for modeling, and the type of analysis possible with rasters.

TYPES OF DATA IN A RASTER

A raster contains one of three types of information: thematic data, spectral data, or pictures.

Thematic data in a raster

A raster can represent a particular phenomena, such as fire, chemical concentration, slope, or elevation. These are typically stored as single-band rasters and often have associated attribute tables.

This raster layer draws the aspect of a terrain.

Aspect is the direction toward which a section of surface is pointing. In this raster, red denotes slopes facing north and yellow shows slopes facing south.

Spectral data in a raster

The most common use of a raster is to present images of the earth acquired through aerial photography or satellite imagery. Specialized cameras can capture the reflectance of light at several or many parts of the spectrum.

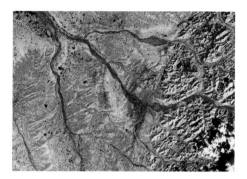

This raster layer shows a multiband spectral raster captured from a satellite imaging system.

In the hands of an imaging scientist, these images can be compared with known spectral signatures of rocks or plants to reveal geological or vegetative structures.

Picture data in a raster

A raster can contain pictures such as scanned maps or building photographs. This type of raster can be single or multiband.

This raster layer shows a picture of a house.

Building pictures can be useful in a property asset application.

DRAWING METHODS FOR RASTER LAYERS

A *raster layer* is a reference to a raster and has an associated drawing method. You can select any attribute of your raster layer and visualize it in a variety of ways.

There are two broad types of raster: single-band and multiband. Some rasters are imaged and some are sampled from other data. The drawing methods for raster layers are described below.

Drawing cells by unique value

A raster can be optionally associated with a table containing attributes for each cell. These cell attributes can describe spatially discrete (or thematic) data such as landuse, soils type, and property ownership.

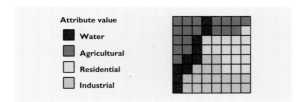

This drawing method is suitable when an attribute exists that already describes a category, type, or classification. The attributes can be descriptive or numeric. Generally, the number of unique values should not exceed 25. Otherwise, it becomes difficult to distinguish between classes.

This raster layer uses unique vegetation codes to draw the distribution of plants for a region.

If a raster has one-bit data, this drawing method can be used with 0 for black and 1 for white to make a monochrome drawing.

Drawing cells classified by graduated colors

Some cell attributes represent a range of numeric values that contain thematic information, such as elevation, slope, pollution contaminants, or population density.

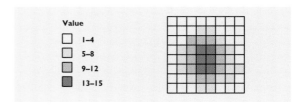

This drawing method lets you define a classification in the same manner as the graduated symbol and graduated color drawing methods for feature layers. You can normalize and exclude attribute values. If a raster has multiple bands, you will select one of those bands for this drawing method.

Once a classification has been made, you can select a color ramp showing each class in a different color. The color ramp you select should provide visual cues that your perception is already accustomed to.

For example, denser concentrations should use bold colors and lighter concentrations should be pale. Temperatures should be blue for cold to red for hot. An elevation map can have two ranges of color, one for elevations above sea level and another for bathymetric elevations.

This raster layer shows discrete elevation ranges displayed with a color ramp.

Drawing cells stretched with graduated colors

Many rasters have continuous data that represents a spectral value, or a calculated value, such as sun illumination angle.

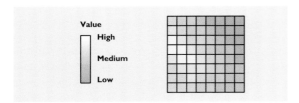

This drawing method is traditionally used for single-band continuous data with a large set of values. It creates high-quality displays of continuous phenomena by rendering them with a continuous gradation of colors.

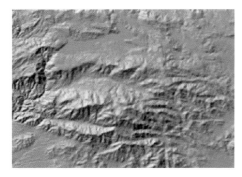

This raster layer shows a raster with sun-illumination angle values stretched from white to black. This type of map is called a hillshade map.

For a raster band, you select the type of *stretch*. A stretch is like a classification, but it describes the rate of change of continuous values.

Some of the available types of stretches include standard deviation, histogram equalize, and minimum–maximum. The stretch calculates and associates high, medium, and low numeric values on a color ramp. Areas of no data can be drawn with a separate color.

Drawing cells using a red-green-blue composite

Rasters that are created for color display are often created with three bands, one each for red, green, and blue.

The types of data collected with these three-band rasters can be satellite imagery, scanned photographs, or any type of picture.

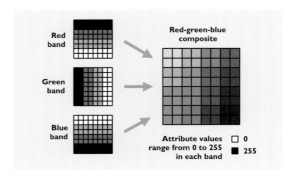

This drawing method uses a collective stretch for the three bands. Areas of no data are drawn with a specified color.

This raster layer shows a red-green-blue color composite satellite image of an urban area adjacent to mountains.

A *triangulated irregular network (TIN)* is an efficient representation of a surface. Rasters are also used for surface modeling, but TINs have the advantage of varying the data density where the surface changes sharply. Areas where the surface is smooth require few points, while areas where the surface is rough require more points.

THE ELEMENTS OF A TIN

A TIN is made from points, each of which have a continuous real value that usually represents elevation. Other types of surface values might be chemical concentration, groundwater level, or precipitation amount.

A triangulation is calculated from these data points and represents a continuous, three-dimensional surface. A triangulation is a nonoverlapping set of triangles, or *faces,* that completely fills a prescribed area.

Because TINs represent a surface with vector features (points, lines, and faces), they can precisely model discontinuities in the surface with breaklines. Examples of breaklines are streams, ridges, and roadways, where the surface slope changes sharply.

One limitation of a TIN is that it cannot represent the rare occurrences of negative slope, such as on vertical cliffs, overhangs, and caves.

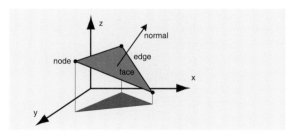

This is a perspective view of one face in three dimensions.

A face defines a plane, a slope, and a slope direction. The *normal* to a face is the perpendicular vector and it is used for calculations such as sun illumination, aspect, and slope.

Chapter 10, "Surface modeling with TINs," has more information on the data structure of TINs, data-sampling considerations, the analytic possibilities

with TIN data models, and a comparison of TINs and rasters for representing surfaces.

DRAWING METHODS FOR TIN LAYERS

A *TIN layer* is a reference to a TIN and an association to one or more TIN drawing methods (or renderers).

You can draw a TIN layer with one or many drawing methods (renderers) that display TIN elements (points, lines, and faces) or visualize qualities of a surface such as elevation, slope, and aspect.

Unlike the drawing methods for rasters and features, a TIN layer allows you to select many drawing methods instead of one. This lets you draw different data elements at once, such as breaklines drawn on top of faces that are colored by elevation.

The following sections describe the drawing methods for TIN layers.

Drawing TIN elements

You can draw the points, lines, and faces in a TIN.

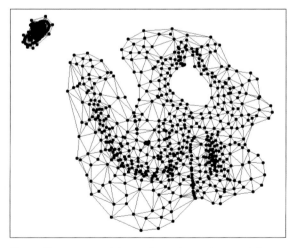

Normally, you would not draw the points and lines in a map presentation, but this option can be useful for inspecting or troubleshooting the point distribution that makes up your TIN.

Drawing TIN faces with hillshading

All of the drawing methods for faces give you the option of *hillshading,* a technique for shading faces to produce realistic views of a terrain.

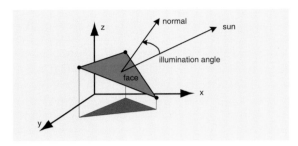

Hillshading works by taking the position of the sun in the sky (which you can control) and calculating the angle between the direction to the sun and the normal to a face.

This angle is used to apply shading on faces that simulates light reflectance off a surface. The brightness of reflected light is proportional to the cosine of the angle between the surface normal and the vector to the light source.

Hillshading creates a realistic three-dimensional image from a two-dimensional display.

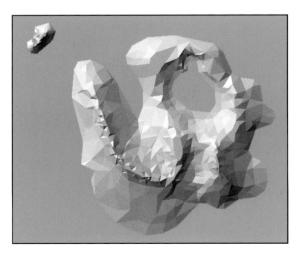

This is a TIN layer with faces drawn with hillshading. The sun is in the northwest at an angle of 30 degrees above the horizon.

Drawing elevation with a graduated color ramp

You can draw the faces in a TIN with colors that show the range of elevations.

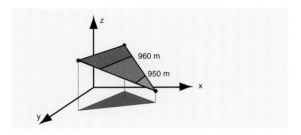

This drawing method interpolates contour lines for each face. A face can have zero, one, or several contour lines that cross it. Each zone between the specified contour interval is drawn with a color from the color ramp.

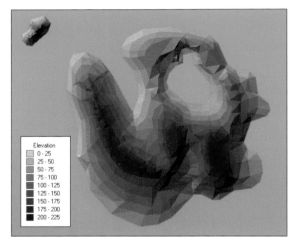

This TIN layer is drawn with elevations rendered on a graduated color ramp and with linear interpolation. Hillshading is also applied.

Drawing aspect with a graduated color ramp

You can draw the cardinal direction, or *aspect,* that each face in a TIN is pointing toward.

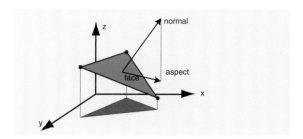

Aspect is the direction on a compass that the normal of a face is pointing to when projected on the plane of the earth. Aspect is measured in degrees. North is 0 degrees, east is 90 degrees, south is 180 degrees, and west is 270 degrees.

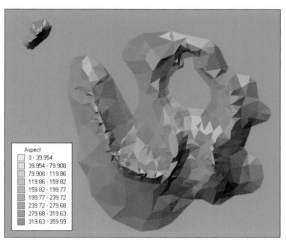

This TIN layer is drawn with aspect rendered on a graduated color ramp. Hillshading is applied.

Drawing slope with a graduated color ramp

You can draw the slope of each face on a surface. This lets you visualize the steep areas of a terrain.

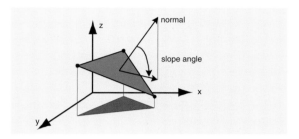

Slope is calculated for each face as the angle between the normal and the plane of the earth. A color ramp is applied to angles between 0 degrees and 90 degrees.

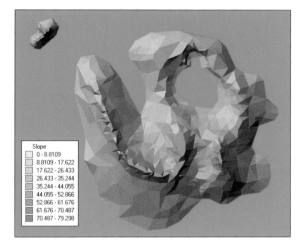

This TIN layer is drawn with slope rendered on a graduated color ramp. Hillshading is applied.

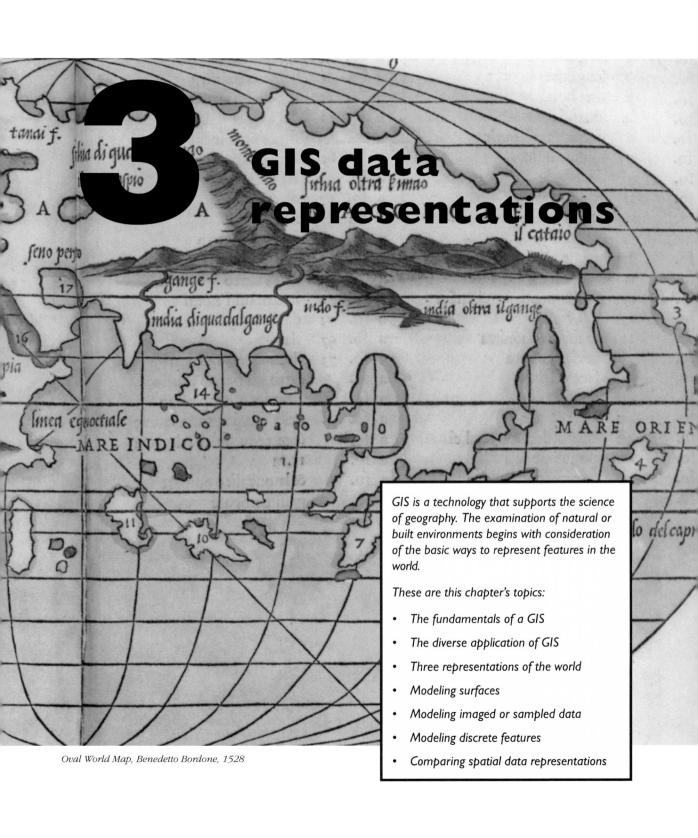

3

GIS data representations

Oval World Map, Benedetto Bordone, 1528

GIS is a technology that supports the science of geography. The examination of natural or built environments begins with consideration of the basic ways to represent features in the world.

These are this chapter's topics:

- The fundamentals of a GIS

- The diverse application of GIS

- Three representations of the world

- Modeling surfaces

- Modeling imaged or sampled data

- Modeling discrete features

- Comparing spatial data representations

This chapter discusses the fundamental concepts you need to understand in order to build superior data models with ArcInfo.

First, a geographic information system (GIS) will be defined. You will review the parts of a GIS, examine how GIS extends a database, and explore the diversity of GIS applications.

Next, you will review some basic concepts of modeling geographic data. You will learn some options for modeling continuous surfaces, discrete features, and imagery. Sometimes, there is more than one reasonable choice for a data model.

THE PARTS OF A GIS

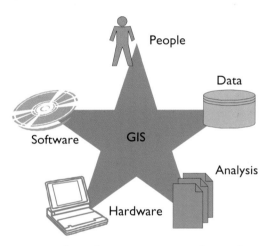

A geographic information system is the combination of skilled persons, spatial and descriptive data, analytic methods, and computer software and hardware—all organized to automate, manage, and deliver information through geographic presentation.

People who build and use GIS

When you design a data model, build a software application, or write user documentation, it is important to be clear on the type of user your work is directed toward.

These are the primary roles that people play in a GIS:

A *map user* is the end consumer of a GIS. This person looks at maps created for a general or specific purpose. All members of the public are map users.

A *map builder* uses map layers from several sources and adds data to make a custom map.

A *map publisher* prints maps. This person is dedicated to high-quality cartographic output.

An *analyst* solves geographic problems, such as chemical dispersion, finding the best route, and site location.

A *data builder* inputs geographic data with several techniques—editing, converting, and data access.

A *database administrator* manages GIS databases and ensures that the GIS operates smoothly.

A *database designer* builds logical data models and implements physical database designs.

A *developer* customizes GIS software to serve the specific needs of an industry.

Data sources for GIS

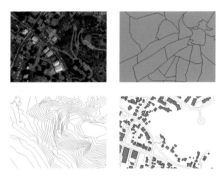

A GIS processes any data that has a spatial component. This information is quite diverse—it can be aerial photographs or satellite imagery, a collection of terrain contours, digital maps of the built environment, or legal records of land ownership.

Geographic data can also reside in some unexpected places—any company that keeps a database of its customers has geographic data. A GIS can calculate the location of any place on earth from a postal address.

Procedures and analysis

The specialists that operate a GIS employ functions, procedures, and judgment. This collective human experience is an indispensible component of a GIS.

Some examples of analytic functions are:

- A science applied in a spatial context, such as hydrology, meteorology, or epidemiology
- Quality assurance procedures to ensure that the data is accurate, consistent, and correct
- Algorithms that solve spatial queries on linear networks or integrated polygon topology
- The knowledge to apply cartographic design principles for excellent map presentation

Computer hardware

Computers come in all sizes, from palm to mainframe. You can purchase GIS software for nearly every type of computer.

With the improvements in network bandwidth, a client–server or n-tier architecture is the preferred configuration for enterprise-scale GIS.

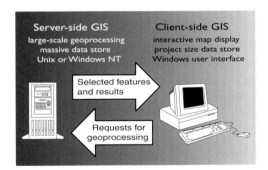

The Internet is joining computers into a global network and is an important way to access data. Another trend is the increasing use of the Global Positioning System (GPS) to locate people in real time.

GIS software: a geographic database

The key idea to grasp about GIS software is that it is, in fact, a geographic database management system. Geodatabases are implemented directly on commercial relational or object-relational database management systems.

The reason for this is to leverage the capabilities of commercial database software, which include data backup, table definition, transaction management, and system administration tools. A GIS extends a relational database so that it can efficiently store geographic data, produce maps, and perform spatial analytic tasks.

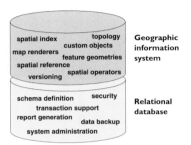

Some of the functions that GIS software adds to a relational database management system are:

- The ability to store the geometric shapes of features directly in a database column.
- A framework to define map layers on data and specify drawing methods; these can be drawn based on attribute values.
- An infrastructure to support the creation of simple and sophisticated maps. Many common map-making tasks are simplified.
- The creation and storage of topologic relationships that exist among features, such as network connectivity and integrated polygon topology.
- A spatial index spanning two dimensions for rapid retrieval of geographic features.
- A set of operators for determining geographic relationships such as proximity, adjacency, overlay, and spatial comparison.
- Many tools to support spatial queries such as network tracing and polygon overlay analysis.
- A work-flow system that allows the editing of geographic data by many users and manages versions.

You can think of a GIS as a spatially enabled database management system. This architecture gives you the best of commercial database technology and sophisticated geographic software.

GIS is being applied in remarkable ways. To understand GIS and see why it matters, it is useful to survey the diverse range of GIS applications.

These are a few descriptions of applications taken from papers submitted at the ESRI user conference.

Agriculture

Satellite images of Brazil showing land use are combined with models of El Niño weather oscillation to predict agricultural effects.

GPS (Global Positioning System) receivers are being applied in real time with portable GIS software to accurately apply chemicals for agricultural production.

In the San Joaquin Valley of California, GIS is used to model nonpoint sources of pollution. The maps produced provide a visual display of soil salinity.

Business geographics

One company used GIS to evaluate how the pending relocation of its corporate office would affect employees' commute to work.

A small company in Quebec facing competitive pressures used GIS to mine its customer databases to identify clusters of customers, enhance productivity of mail promotions, and improve client retention.

A foundation in San Francisco uses GIS to assist small businesses in finding commercial space with desirable business, economic, demographic, and transportation attributes.

Defense and intelligence

The U.S. Air Force uses GIS technology to manage, maintain, and visualize millions of climatological records.

The Swedish armed forces have done extensive work on sophisticated symbolizing of military and civilian facilities to improve military planning.

The Canadian Army has customized GIS software to integrate it with a land force command system.

Ecology and conservation

Colombia is building a GIS database to prioritize which lands should be set aside for the national park service.

In Kenya, a GIS revealed that large mammals in the savanna dispersed during the wet season and concentrated in a basin during the dry season. Understanding seasonal migration patterns is important in managing water access for wildlife and livestock.

GIS is being applied on California's Santa Catalina Island to evaluate the ecological costs and benefits of dirt roads. Roads pose an environmental dilemma because they provide access for ecological management but also interrupt the ecological landscape.

Electric and gas (AM/FM)

Beirut is analyzing its power circuits to minimize losses and to improve voltage levels. GIS is modeling scenarios of device placement for optimal electrical benefit.

Public Service of New Mexico is using GIS to manage the construction, operation, and maintenance of 2,500 miles of power transmission. A prime concern is preventing environmentally damaging activities.

The Danish Energy Agency is building a database on the energy usage of every building in that country. This information will be used for planning energy plants and designing distribution systems.

Emergency management and public safety

In 1997, the Cassini spacecraft was launched to explore Saturn. GIS was used to evaluate the risk of an accident with the plutonium generators on board.

The Italian National Seismic Survey is building an integrated information system to produce real-time tabular reports and operational maps in the event of a major earthquake.

Environmental management

In Korea, land zoning in national parks is being analyzed with the criteria of scenic quality, elevation, slope, and natural state. It was found that some parks were not correctly zoned.

A large dam is being constructed in Turkey. GIS is being used for a complete evaluation of its effects on irrigation, hydropower, health, mining, education, tourism, and telecommunications.

In Bavaria, ecological balance models are combined with GIS software to provide tools for environmental management. This information is disseminated through the Internet.

Federal government systems

The Tennessee Valley Authority has built a land information system to help administer land records, natural and cultural resources, land-use planning, and compliance with laws and executive orders.

The U.S. National Oceanic and Atmospheric Administration is building a tool to collect metadata such as bounding coordinates, map projections, and attribution information.

Forestry

The construction and use of roads in a forested basin can contribute significantly to sediment deposition. A forestry company is building a road sediment model to establish a maintenance plan.

The U.S. Fish and Wildlife Service has established guidelines for managing forests where the red-cockaded woodpecker, an endangered species, is found. GIS is used to calculate colony areas and foraging zones.

Health care

The State of California is mandating that county governments address cultural and ethnic issues for outpatient health care. GIS is being used to present geographic, socioeconomic, demographic, and health care utilization data.

A university researcher is using GIS to analyze the epidemiology of rare diseases and estimating an individual's exposure to environmental risk factors.

In Colorado, the percentage of low-birthweight babies exceeds the national average. GIS is being used to examine factors such as age, race, education, elevation, and access to public health programs.

Education

An educational agency is using GIS to help students discover geography and foster critical thinking and inquiry.

A high school is incorporating GIS in its curriculum to teach students a "sense of place" by showing them how their personal actions have relevance on a global scale.

Mining and geosciences

GIS is used in West Virginia to monitor acid mine drainage on surface waters. Elevations, hydrology, mined areas, and water quality data are combined.

A mining services company is using GIS to create three-dimensional databases for nuclear waste repositories, mineral exploration programs, and groundwater monitoring purposes.

Oceanography, coastal zone, marine resources

The U.S. Naval Oceanographic Office is using remotely sensed sea temperature data to study oceanic fronts and eddies.

In Washington state, a GIS is mapping the current shoreline, calculating change rates, and projecting shoreline erosion hazards.

Real estate

Habitat for Humanity, an organization building houses for low-income families, uses GIS to analyze a proposed subdivision and create a plan that preserves most of the existing trees.

A realty company is using GIS for site selection for multisite users. Factors considered are access, visibility, zoning, and entitlement processes.

Remote sensing and imagery

A digital imagery company is using georeferenced airborne sensors to create real-time spatial data. Images are sent to ground stations and are fused, reformatted, and subjected to automatic feature extraction.

State and local government

In Qatar, television cameras are being inserted into water and sewer networks to create video records of pipe conditions. These images are integrated within a GIS and give operators information for maintenance.

The new Denver International Airport is located in a rural area. GIS is being applied to develop scenarios of land-use patterns over the next five, 10 and 15 years.

In the Ukraine, political changes have ushered in a wave of land reform. Lack of accurate records has hampered the creation of an accurate cadastre, so a new land registration system is being developed based on high-resolution satellite imagery and innovative software techniques.

Telecommunications

In Colombia, the fiber-optic trunk network is being captured in a GIS database with a representation of each of the network's element features.

In Indonesia, GIS is employed to manage radio telephony by studying radio station placement, the demographics of a customer area, and the maintenance of equipment.

A telecommunications consulting firm is using data on land use and land cover to predict signal attenuation for wireless communication systems.

Transportation

In Korea, a GIS monitors real-time traffic conditions to mitigate traffic bottlenecks on freeways.

The State of Georgia applies GIS technology to manage roadway pavement. A study was made of road segment ratings based on load cracking.

Water distribution and resources

Population growth and agricultural expansion in Egypt are placing demands on water management. A government ministry is building a system to manage the Nile River channel, canals, drains, and pumps.

In Florida, a hydraulic computer model is used to reduce sanitary sewer overflows. When major rainstorms come, satellite imagery is used to estimate rainfalls and assist in the operation of sewer pump stations.

In Canada, a hydrodynamic/pollutant transport model has been built to simulate the effects of multiple pollution sources under different conditions.

SUMMARY OF GIS APPLICATIONS

These applications prove the diversity of GIS solutions. It is always surprising to discover how widely ranging the uses of GIS technology are. Common characteristics throughout these applications include:

- Frequently, GIS is integrated with other applications to perform geographic and scientific analysis. It is important that GIS data be structured and stored in a way that allows for distributed data access.

- An open data architecture has considerable importance in facilitating the integration of geographic data with other data, such as real-time data, imagery, and corporate databases.

- While printed maps are still the most common presentation of geographic data, Internet map access and dynamic map applications are becoming increasingly important for decision making. Interactive access invites more sophisticated data models to support rich queries and analysis.

- Selecting the right data structure is important to enable the kind of analysis you wish to perform. These applications illustrate many skillful applications of modeling the world as a continuous surface, as a raster grid, or as sets of discrete features in vector format.

The applications just reviewed show patterns of historic use, capture the current state of the natural and built environments, and predict changes in the world based on weather, human activity, or geophysical events. In every application, a decision was made concerning applying physical datasets to serve logical data models.

DATA REPRESENTATION MODELS

With a GIS, you can model data in three basic ways: as a collection of discrete features in vector format, as a grid of cells with spectral or attribute data, or as a set of triangulated points modeling a surface.

Modeling with vector data

Vector data represents features as points, lines, and polygons and is best applied to discrete objects with defined shapes and boundaries.

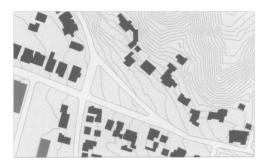

Features have a precise shape and position, attributes and metadata, and useful behavior.

Modeling with raster data

Raster data represents imaged or continuous data. Each cell (or pixel) in a raster is a measured quantity.

The most common source for a raster dataset is a satellite image or aerial photograph. A raster dataset can also be a photograph of a feature, such as a building.

Raster datasets excel in storing and working with continuous data, such as elevation, water table, pollution concentration, and ambient noise level.

Modeling with triangulated data

A TIN is a useful and efficient way to capture the surface of a piece of land.

TINs support perspective views. You can drape a photographic image on top of a TIN for a photorealistic terrain display. TINs are particularly useful for modeling watersheds, visibility, line-of-sight, slope, aspect, ridges and rivers, and volumetrics.

TINs can model points, lines, and polygons. A triangulation is made of many *mass points,* each an x,y,z tuple. *Breaklines* represent streams, ridges, and other linear discontinuities. *Exclusion areas* represent polygons with same elevation, such as lakes or project boundaries. Contour maps can be generated from a TIN, using linear interpolation or a smoothing algorithm.

Implementing data representation models

A geodatabase implements the vector data representation with *feature datasets* and *feature classes,* the raster data representation with *raster datasets,* and the triangulated data representation with *triangulated irregular networks (TINs).*

A GIS can model a surface in three general ways: as a surface raster, as contour lines, or as a triangulated irregular network.

Each approach has merit, but the triangulated irregular network has special analytic powers and the surface raster can also perform interesting analysis.

SURFACE RASTER

Some terrain data comes in the form of a uniform grid with elevation values. An example is the Digital Elevation Model (DEM) data product from the United States Geological Survey.

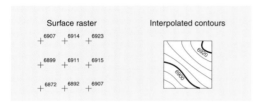

A raster dataset can represent point elevations spaced at regular intervals. Each cell in the raster has an associated elevation value.

From a raster dataset with elevations, the elevation for any point on a surface can be estimated and a set of contours can be derived.

The advantages of raster datasets are:

- The conceptual model of raster datasets is simple. Data storage is very compact.

- The raster model has well-established algorithms to process raster data.

- Elevation data in raster format is relatively abundant and inexpensive to obtain.

The disadvantages of raster datasets are:

- The rigid grid structure does not conform to the variability of terrain.

- The original data is not maintained when it is interpolated to a regularly spaced grid.

- Linear features cannot be represented well for many applications.

CONTOUR LINES

Contour lines can be used to represent surfaces. A contour is a line following an equal elevation value. Contours are the most accessible source of terrain information for most map users.

Contours are good for human interpretation. Closely spaced contours are a clear visual cue that the local terrain is steep. A sharp angle in a contour is a clue of a stream or ridge line. You can get a sense of the "lay of the land" by reading contours on a map.

However, contours are generally poor for computer surface models. The collection of all points on contours does not make a good dataset for surfaces. It is difficult to remove data artifacts introduced from converting contours to rasters or TINs. Converting contours is usually a last resort for building a surface model.

You can make a perspective view or perform surface analysis of contours only after they have been converted to a raster or a TIN.

TRIANGULATED IRREGULAR NETWORKS

A *triangulated irregular network* (TIN) is an efficient and accurate model for representing continuous surfaces. TIN software includes many functions that analyze surfaces.

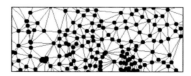

A TIN dataset is built in this way:

1. Collect a set of points with x,y,z coordinates through photogrammetric instruments, GPS data collection, or other means. Collect breaklines where the shape of the surface changes sharply. Collect exclusion areas for features such as lakes.

2. From this point data, GIS software creates an optimal network of triangles, called a Delaunay triangulation. In a TIN, each triangle is created to be as close to equilateral as possible.

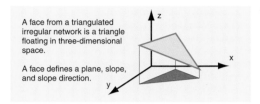

A face from a triangulated irregular network is a triangle floating in three-dimensional space.

A face defines a plane, slope, and slope direction.

3. Each triangle forms a face with a gradient slope.

From a TIN, an elevation can be calculated for any point with x and y values by first locating the triangle and then interpolating the height inside it.

A TIN is efficient because the point density on any part of the surface can be proportional to the variation in terrain. A flat plain suffices with a low point density and mountainous terrain requires a high point density, especially where the surface changes abruptly.

Elements of a TIN

A TIN can represent points, lines, and polygons.

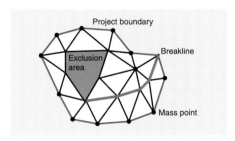

Mass points are observed spot elevations with an x,y,z coordinate triplet. They can be collected with photogrammetric instruments, remote sensors, or data conversion.

Breaklines delineate where the terrain has a sharp discontinuity in its surface. Examples of features modeled as breaklines are streams, ridges, and the edges of building pads or other areas graded by machinery.

An area of exclusion delimits an area of equal elevation. These are most commonly lakes.

Also, a project boundary can exclude the surface outside an area of interest. This can be important when you are calculating volumes.

Displaying surfaces with a TIN

There are several ways to visualize the surface represented by a TIN. You can draw a TIN on a planimetric (two-dimensional) map with colors representing elevation, slope, and aspect.

With three-dimensional extension software to ArcInfo, you can display perspective views of a surface with draped images, contours, grid lines, or other features.

Analysis with a TIN

TIN software includes various analytic tasks on a surface. Some of the tasks are:

- Calculate the elevation, slope, and aspect (the compass direction of slope) for any point within the surface.

- Generate contours by linear or quintic (smooth) interpolation across the triangulation.

- Determine a range of elevations for a surface.

- Summarize statistics for a surface, such as volume against a reference plane, mean slope, area, and perimeter.

- Create vertical profile displays along alignments on the surface.

- Perform volume calculations for roadway projects, so that the volume excavated in one area equals the volume deposited in another.

- Analyze which areas of a surface are visible from a point.

Image data is collected by satellite systems or aerial photography. Because this is by far the least expensive way to collect vast quantities of geographic data, images are an important component of many GISs.

RASTER DATASETS

Raster data can be used as a backdrop to a map display, as a source for feature extraction, for gridded surface models, or for modeling proximal geographic functions such as dispersion. GIS software can rapidly overlay stacked raster datasets.

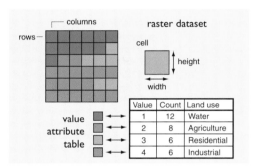

A raster dataset stores a two-dimensional matrix with sampled values for each cell. Each cell has the same width and height.

The geographic coordinate of the upper-left corner of the grid, together with the cell size and number of grid rows and columns, uniquely defines the spatial extent of the raster dataset.

Cell values for raster datasets can be integer or floating numbers. Some representative types of values for raster cells include:

- Light reflectance (albedo) in a photograph.

- Light intensity at a specific part of the spectrum in a satellite image.

- A derived attribute, such as land-use type, or a feature type, such as a building or street.

- A z value, such as elevation or concentration.

A value attribute table (VAT) can be optionally associated with a raster dataset. This table keeps track of your value classification. You can add custom attributes by adding more columns.

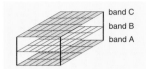

Raster datasets can have one or many bands. Each band in a raster dataset has an identical grid layout but represents a different attribute. The most common use of multiple bands is to represent the multispectral data captured by satellite imagery.

Raster datasets as feature attributes

Not all raster datasets have a geographic reference. An image can be used as an attribute to a feature.

If you are building a GIS to sell homes, you may want an Internet application where the prospective buyer is shown a map with symbols for each home for sale. The buyer can click a symbol to display an image, facts about the house, and the price.

Other examples of images as feature attributes are:

- Scanned documents, such as permits or deeds.
- Field data forms associated with locations.
- Blueprints or schematic diagrams of floorplans.

Representing points, lines, and polygons

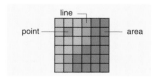

Points in a raster dataset can be represented by one or a few contiguous cells. Lines can be represented by a series of cells that have a width of one or a few cells. Polygons can be represented by a range of cells. Although you can visually identify points, lines, and polygons in a raster dataset, it is best to do a raster-to-vector conversion if you want to interact with features.

Converting raster datasets

Raster datasets can be generated with ease, yet the features they depict are sometimes more useful in another type of dataset. An example is converting a photograph of buildings into a feature dataset containing buildings with polygon shape.

The resolution of the raster dataset strongly influences the accuracy of converted vectors.

Raster analysis

GIS software for raster datasets comes equipped with a powerful set of operations. These are a few:

- Spatial transformations. A raster dataset can be moved, bent, or stretched to fit the true location. It can also be projected on a coordinate system. Rubber sheeting locally adjusts rasters to fit user-defined vectors. Polynomial transformations apply global equations to fit grids to user-defined vectors.

Land use + Water = Habitat

- Spatial coincidence. Modeling characteristics of locations, such as assessing the suitability for

some type of land development like the optimal location of a new road, or estimating land values.

- Proximity. Modeling the distance to other geographic phenomena. This distance can be measured as straight Euclidean distance or an abstraction, such as travel time.

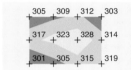

- Surface analysis. Finding the qualities of continuous surfaces, such as elevation, noise, or pollution concentration. You can calculate slope and aspect from a land surface or determine the noise level in the vicinity of an airport.

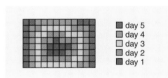

- Dispersion. Modeling the movement of phenomena, such as simulating the spreading of fire or predicting the movement of an oil spill.

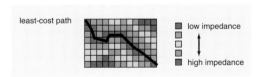

- Least-cost path analysis. You can calculate the shortest path across the surface based on any desired impedance values.

Geographic features are located at or near the surface of the earth. Geographic features can occur naturally (rivers, vegetation), can be constructions (roads, pipelines, buildings), and can be subdivisions of land (counties, land parcels, political divisions).

Maps model the world with points, lines, and polygons.

- Points represent geographic features too small to be depicted as lines or areas.

- Lines represent geographic features too narrow to be depicted as areas.

- Polygons represent sizeable continuous geographic features.

An x,y (Cartesian) coordinate system references real-world locations.

FEATURE DATASETS

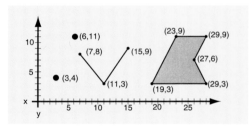

In feature datasets, each location is recorded as a simple x,y coordinate. Points are recorded as a single coordinate. Lines are recorded as a series of ordered x,y coordinates. Polygons are recorded as a series of x,y coordinates defining line segments that enclose an area.

Point features

Points represent geographic features that have no area or length, or features that are too small for their boundaries to be apparent for a given map scale.

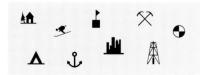

Line features

Linear features represent objects that have length but no area, or features whose shapes are very narrow at a given map scale.

Polygon features

Polygon features are used to represent areas such as states, counties, census tracts, sales territories, soil units, parcels, and land-use zones.

Polygons enclose areas that meet a user-specified set of common characteristics for the phenomena being represented.

How maps convey descriptive information

Maps present descriptive information about geographic features using symbols and labels.

Here are some common ways that maps present attribute information about the geographic features they represent:

- Roads are drawn with various widths, patterns, and colors to represent different road classes or other attributes.

- Streams and water bodies are typically drawn in blue to indicate water.

- Special symbols denote specific features such as railways and airports.

- City streets are labeled with names and often address ranges.

- Special buildings are labeled with their names or functions.

FEATURES, NETWORKS, AND TOPOLOGY

Features can have three basic roles with respect to one another: simple, networked, and topological.

Simple features

Features can be simple, with no explicit connections or topological associations to other features.

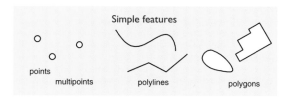

Network features

Features can be connected in a network.

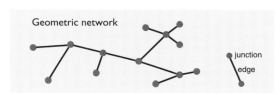

A network contains *edges* that have *nodes* at their endpoints. A node can be connected to one or many edges. This assemblage of edges and nodes is called a *geometric network*.

Shared edges in a topology

The topological elements of features can be derived and edited.

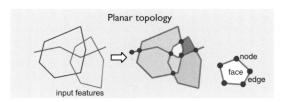

In the ArcMap editor, you can specify a set of features and create a *planar topology,* which is a set of topological primitives: nodes, edges, and faces.

When you edit a node, the connecting edges rubberband. When you edit an edge, you are modifying the shape of two faces at once.

FEATURES AND CARTOGRAPHY

Features are geographic objects in the context of a map. And a map has a scale, which determines the feature's dimension: point, line, or polygon.

Buildings can be drawn as polygons at large scale and points at small scale.

A stand of trees might be drawn individually at large scale, but a forest is drawn as polygons that bound trees above a certain density.

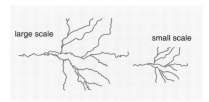

A stream system at large scale will have many shape points and minor branches. At small scale, the line detail is filtered and minor streams are removed.

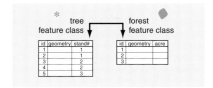

If you need to change the feature dimension at varying scales, you can set up a database relate to another feature class. In this case, trees are associated with a forest stand. When you draw a map, the scale determines which set of features is drawn.

	Vector data representation	Raster data representation	Triangulated data representation
Focus of model	Vector data is focused on modeling discrete features with precise shapes and boundaries.	Raster data is focused on modeling continuous phenomena and images of the earth.	Triangulated data is focused on an efficient representation of a surface that can represent elevation or other quality, such as concentration.
Sources of data	Compiled from aerial photography Collected from GPS receivers Digitized from map manuscripts Sketched on top of raster display Vectorized from raster data Contours from triangulation Reduced from survey field data Imported from CAD drawings	Photographed from an airplane Imaged from a satellite Converted from a triangulation Rasterized from vector data Scanned blueprints, photographs	Compiled from aerial photography Collected from GPS receivers Imported points with elevations Converted from vector contours
Spatial storage	Points stored as x,y coordinates. Lines stored as paths of connected x,y coordinates. Polygons stored as closed paths.	From a coordinate in the lower-left corner of the raster and cell height and width, each cell is located by its row and column position.	Each node in a triangle face has an x,y coordinate value.
Feature representation	Points represent small features. Lines represent features with a length but small width. Polygons represent features that span an area.	Point features are represented by a single cell. Line features are represented by a series of adjacent cells with common value. Polygon features are represented by a region of cells with common value.	Point z values determine the shape of a surface. Breaklines define changes in the surface such as ridges or streams. Areas of exclusion define polygons with the same elevation.
Topological associations	Line topology keeps track of which lines are connected to a node. Polygon topology keeps track of which polygons are to the right and left sides of a line.	Neighboring cells can be quickly located by incrementing and decrementing row and column values.	Each triangle is associated with its neighboring triangles.
Geographic analysis	Topological map overlay Buffer generation and proximity Polygon dissolve and overlay Spatial and logical query Address geocoding Network analysis	Spatial coincidence Proximity Surface analysis Dispersion Least-cost path	Elevation, slope, aspect calculations Contour derivation from surface Volume calculations Vertical profiles on alignments Viewshed analysis
Cartographic output	Vector data is best for drawing the precise shape and position of features. It is not well suited for continuous phenomena or features with indistinct boundaries.	Raster data is best for presenting images and continuous features with gradually varying attributes. It is not generally well suited for drawing point and line features.	Triangulated data is best for rich presentation of surfaces. This data can be viewed by using color to show elevation, slope, or aspect or in a three-dimensional perspective.

In summary, there are three basic representations of spatial data: vector, raster, and triangulated. Each of these representations has merits and is well suited for a particular class of geographic analysis and cartographic output.

These spatial data representations are not exclusive; your geodatabases can contain all three for the map uses for which they are best suited. A map can display any or all of these spatial data representations.

Often, raster data is displayed as a background layer to vector data. This provides a photo-realistic context to the vector layers on which you might be performing engineering or analysis.

Triangulated data is also sometimes drawn as a background layer to vector data to provide a visualization of the shape of the earth's surface.

CHOOSING A SPATIAL DATA REPRESENTATION

You must consider many issues when choosing a spatial data representation. Often, the choice is clear and guided by the available data and the analytical tasks you need to perform. Sometimes, it is not so obvious which data representation is best.

Surfaces are a good example; two robust ways to represent a surface are as raster data or as triangulated data. A choice requires more study.

The following are a few considerations for choosing a spatial data representation.

Is the focus on features or location?

If you are modeling distinct objects with attributes and behavior, the vector data representation is superior.

If you are modeling continuous objects or phenomena characterized by an attribute at a location, you should choose between raster or triangulated data.

Raster data models an area with uniform sampling of attributes in a regular grid. Triangulated data models an area with points and values sampled at a variable density.

What data is readily available?

A major influence on your selection of a data representation is what data is already available.

An early step in the design of your GIS is a survey of all the geographic data already available. When you find the data that is most suitable, you will make a judgment on whether that data is sufficient or whether you will need to create new data by other means such as aerial photography, GPS data collection, or digitization.

Sometimes, you might choose to convert existing data from one representation to another. For example, the best source for electric transmission lines might be scanned maps in raster format. To perform electrical analysis or environmental studies, you may find it necessary to convert that raster data into vector data. You will weigh the cost and quality of output of this raster-to-vector conversion by another means of data collection.

What is the required precision for locating features?

If you need to locate features with significant precision, you should choose vector data representation. Feature identification and selection is easier with vector data, and precise coordinate values are stored.

Determining locations of features in raster data is constrained by the dimensions of each cell. In triangulated data, only the locations of points and breaklines are well defined. The locations of features and their shapes are generally indistinct in raster and triangulated data.

What types of features are required?

If you are modeling large features with values that vary, change with time, or have indistinct boundaries, the raster data representation is usually best. An example is the modeling of a fire over time or the dispersion of groundwater contaminants.

If you are modeling features that characterize the shape of the earth's surface, such as mountain peaks, ridge lines, or stream lines, the triangulated data representation is often best.

Some natural features are better represented with vector data. An example is a river system. If you are displaying rivers as a background layer on a map or modeling ship traffic on a river as part of a broader transportation analysis, you will probably choose the vector data representation.

If you are modeling man-made features, the vector data representation is most often best. Man-made features have well-defined shapes that are characterized by straight lines and circular arcs. Also, man-made features are often located with survey-level precision.

What type of topological association is desired?

Some objects are nontopological and can be freely placed in a geographic area. For example, an area defining a wildlife habitat is arbitrary, ill-defined, and overlaps other habitats, and so it does not have a topological relationship with other features.

Also, many objects are primarily stored in a GIS for the purpose of background display on a map, so it is usually not necessary to store them in a topological format. If roads are a background layer in your GIS, they will probably be simple features. If roads are part of an analysis of a transportation system, they should be topological features.

A GIS can have networks and topologies and these are captured within the vector data representation. Networks represent roads, rivers, and utilities. Topologies represent collections of areas where each point in an area is covered by exactly one polygon.

What type of analysis is required?

If you are analyzing a surface, the triangulated data representation supports the broadest array of analytic functions. However, the raster data representation also represents some surface-modeling functions.

The triangulated data representation supports volume calculations between two surfaces representing undeveloped and developed areas, what area is visible from a point in space, the determination of elevation, slope, and aspect for any point on a surface, and the generation of vertical profiles of an alignment, such as road or utility.

If you are analyzing dispersion of an indeterminate feature over time, such as a plume of pollution, then you should select the raster data representation. The raster data representation also supports the determination of proximity to features, least-cost paths, and rapid overlay of rasters for suitability analysis.

If you are locating optimal locations for placing a business or performing a service, studying flows through a network, managing land records, referencing postal addresses to a location on a map, or querying features on a map, you should choose the vector data representation.

The vector data representation allows analysis that is based on spatial relationships such as proximity and adjacency, and topological relationships such as upstream and connected.

What types of maps are to be produced?

The type and quality of the desired cartographic presentation also guides which spatial data representation is recommended.

The raster and triangulation data representations produce attractive maps of areas with varying attribute values. The vector data representation makes maps with fine detail for features.

Cartographic considerations will further guide whether points, lines, or polygons for vector data representation are best. For example, the map scale will guide whether buildings should be represented as points or polygons, or rivers as lines or polygons.

CONCLUSION

ArcInfo provides a rich infrastructure for the three fundamental representations of geography. The next chapter reveals how geographic data is structured and presented in the ArcInfo applications.

4

The structure of geographic data

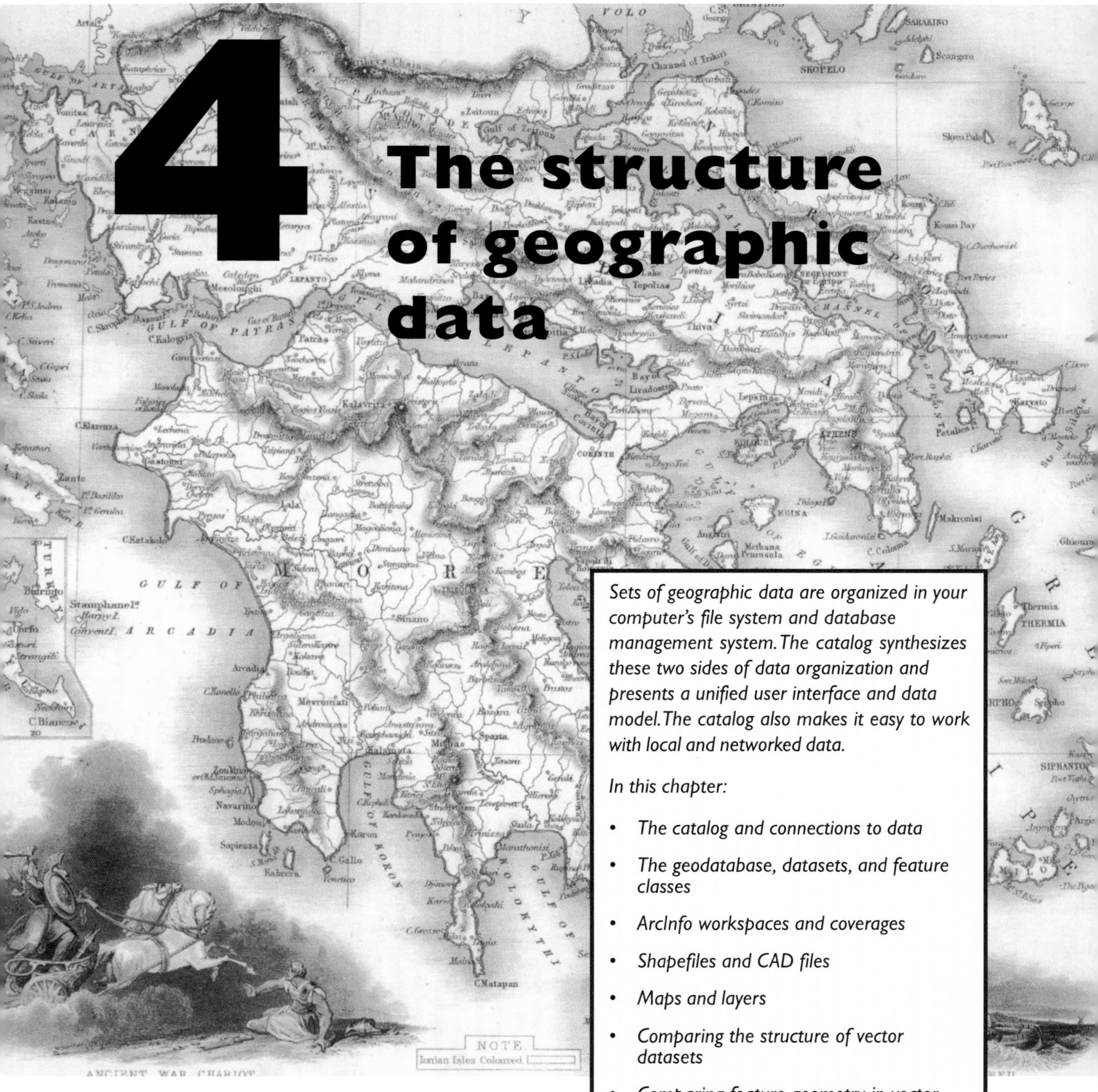

Sets of geographic data are organized in your computer's file system and database management system. The catalog synthesizes these two sides of data organization and presents a unified user interface and data model. The catalog also makes it easy to work with local and networked data.

In this chapter:

- The catalog and connections to data

- The geodatabase, datasets, and feature classes

- ArcInfo workspaces and coverages

- Shapefiles and CAD files

- Maps and layers

- Comparing the structure of vector datasets

- Comparing feature geometry in vector datasets

Ionian Isles and Greece, John Rapkin, 1851.

Your desktop computer has all sorts of data organized into folders, documents, spreadsheets, and databases. You use documents for letters and reports, spreadsheets for expense reports, and databases to keep inventories of customers and products. And you organize files into a folder hierarchy that makes sense; you might organize them by clients, projects, time span, or any meaningful association.

Likewise, a geographic information system (GIS) manages data in a hierarchy of folders, files, and geodatabases. The primary types of geographic data —vector, raster, TIN, locations—can be contained within geodatabases or files.

Your geographic data can be hosted in a single-user geodatabase on your computer's disk or a multiuser geodatabase hosted on a database server. And you can structure your geodatabases and folders to reflect project areas, thematic groupings, department organization, or other ordering.

THE CATALOG

ArcCatalog is the ArcInfo application that lets you explore, access, manage, and build your geographic data. It presents your geographic data in a manner similar to Microsoft's Windows Explorer.

The items you see in the catalog represent data objects such as geodatabases and feature classes, map objects such as maps and layers, and ancillary objects such as styles and coordinate systems.

The collection of connections to geographic data is called a *catalog*. A catalog provides you with a seamless view of geographic data—file-based data and personal databases are located within a recognizable tree hierarchy. A catalog can also drill into relational databases and reveal some of their internal structure, particularly the tables that store geographic data.

The catalog reveals the structure of geographic data through special icons that communicate the role of the various elements of your geographic database. Some catalog items represent folders and files in your Windows file system. Other items represent collections of features and objects within geodatabases. Certain items are references to

geodatabases and relational databases accessed across a network.

Some tasks you can perform with a catalog include:

- Creating and formatting new data
- Searching for data
- Assessing geographic extent and suitability of data
- Documenting the provenance and quality of data
- Launching GIS operations
- Publishing data for widespread access

Single-user and multiuser geodatabases

Geodatabases come in two basic variants—personal geodatabases using Microsoft Access .mdb files and multiuser geodatabases served through ESRI's Arc Spatial Database Engine (ArcSDE) to one of the leading relational database management systems.

These two variants are functionally identical except that multiuser geodatabases support versioning, which allows multiple users to access and edit the common geographic database. Versioning is discussed further in chapter 7, "Managing Work Flow with Versions."

Folder connections and database connections

Folder connections and database connections give you a consistent and unified view of all your data.

A folder connection lets you access data on your local drives or shared drives on networked computers.

A database connection contains the specifications for accessing a database: server or IP address, instance or TCP port information, and account user name and password. You can access geographic data in a relational database management system through ArcSDE or you can access nonspatial attribute data through an ODBC (Open Database Connectivity) driver.

Once you connect to a remote multiuser geodatabase and expand its tree nodes in the catalog, you will see exactly the same structure of constituent data objects on folder connections.

The catalog, folders, and connections

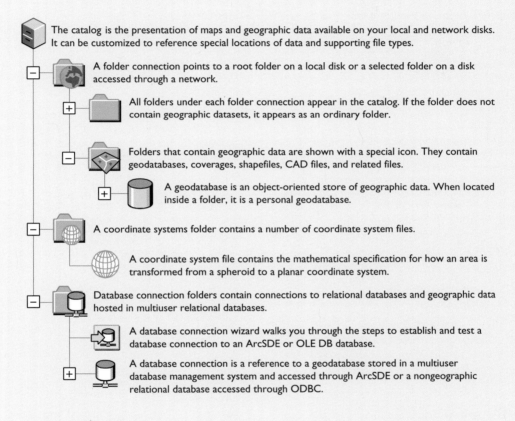

The catalog is the presentation of maps and geographic data available on your local and network disks. It can be customized to reference special locations of data and supporting file types.

A folder connection points to a root folder on a local disk or a selected folder on a disk accessed through a network.

All folders under each folder connection appear in the catalog. If the folder does not contain geographic datasets, it appears as an ordinary folder.

Folders that contain geographic data are shown with a special icon. They contain geodatabases, coverages, shapefiles, CAD files, and related files.

A geodatabase is an object-oriented store of geographic data. When located inside a folder, it is a personal geodatabase.

A coordinate systems folder contains a number of coordinate system files.

A coordinate system file contains the mathematical specification for how an area is transformed from a spheroid to a planar coordinate system.

Database connection folders contain connections to relational databases and geographic data hosted in multiuser relational databases.

A database connection wizard walks you through the steps to establish and test a database connection to an ArcSDE or OLE DB database.

A database connection is a reference to a geodatabase stored in a multiuser database management system and accessed through ArcSDE or a nongeographic relational database accessed through ODBC.

When you design and implement your geographic data model, you have considerable discretion to design each level of your file management system and database schema. The catalog can adapt to your existing organization of data or you can devise a new structure optimized for access and administration.

ORGANIZING GEOGRAPHIC DATA

Geographic data is organized into a hierarchy of data objects. In your data design process, you can organize your data by work groups, thematic type, common spatial extent and coordinate system, or topological associations.

Geodatabases

A *geodatabase* is the top-level unit of geographic data. It is a collection of datasets, feature classes, object classes, and relationship classes.

Your sum aggregate of geographic data can span one, several, or many geodatabases. Geodatabases are usually organized into broad categories of data such as land base, transportation, environment, and utility infrastructure.

Geodatabases manage seamless geographic data. There is no partitioning of a geographic area into tiled units. Rather, geodatabases use effective spatial indexing for continuous representation of an extent.

Personal geodatabases can represent small- to medium-sized datasets. Very large datasets can be efficiently handled with an enterprise ArcSDE implementation.

Geographic datasets

There are three general types of geographic data models: vector, raster, and triangulation. In the geodatabase, they are implemented by three types of geographic datasets: the feature dataset, the raster dataset, and the TIN dataset.

A *feature dataset* is a collection of feature classes that share a common coordinate system. You may choose to organize simple feature classes inside or outside of feature datasets, but topological feature classes must be contained within a feature dataset to ensure a common coordinate system.

A *raster dataset* can either be a simple dataset or a compound dataset with multiple bands for distinct spectral or categorical values.

A *TIN dataset* contains a set of triangles that exactly span an area with a z value for each node that represents some type of surface.

Object classes

An *object class* is a table in a geodatabase with which you can associate behavior. Object classes keep descriptive information about objects that are related to geographic features, but are not features on a map.

An example of an object class is owners of land parcels. You can establish a database join between a polygon feature class for land parcels and an object class for owners.

Feature classes and topology

A *feature class* is a collection of features with the same type of geometry: point, line, or polygon. You can think of two categories of feature classes—simple and topological.

Simple feature classes contain points, lines, polygons, or annotation without any topological associations among them. That is, points in one feature class may be coincident with, but distinct from, the endpoints of lines in another feature class. These features can be edited independently of each other.

Topological feature classes are bound within a *graph,* which is an object that binds a set of feature classes that comprise an integrated topological unit. ArcInfo 8 introduces the first type of graph in a geodatabase—geometric networks.

Relationship classes

A *relationship class* is a table that stores relationships between features or objects in two feature classes or tables. Relationships model dependencies between objects.

With relationships, you can control what happens to an object when its related object is removed or changed.

The catalog's view of a geodatabase

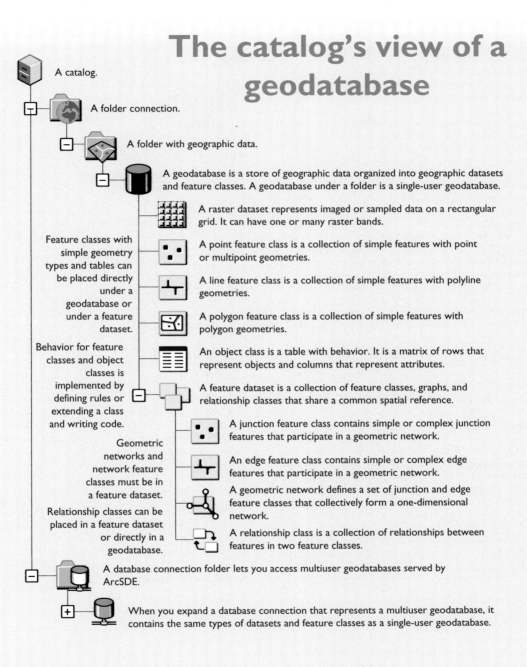

A catalog.

A folder connection.

A folder with geographic data.

A geodatabase is a store of geographic data organized into geographic datasets and feature classes. A geodatabase under a folder is a single-user geodatabase.

A raster dataset represents imaged or sampled data on a rectangular grid. It can have one or many raster bands.

Feature classes with simple geometry types and tables can be placed directly under a geodatabase or under a feature dataset.

A point feature class is a collection of simple features with point or multipoint geometries.

A line feature class is a collection of simple features with polyline geometries.

A polygon feature class is a collection of simple features with polygon geometries.

Behavior for feature classes and object classes is implemented by defining rules or extending a class and writing code.

An object class is a table with behavior. It is a matrix of rows that represent objects and columns that represent attributes.

A feature dataset is a collection of feature classes, graphs, and relationship classes that share a common spatial reference.

A junction feature class contains simple or complex junction features that participate in a geometric network.

Geometric networks and network feature classes must be in a feature dataset.

An edge feature class contains simple or complex edge features that participate in a geometric network.

A geometric network defines a set of junction and edge feature classes that collectively form a one-dimensional network.

Relationship classes can be placed in a feature dataset or directly in a geodatabase.

A relationship class is a collection of relationships between features in two feature classes.

A database connection folder lets you access multiuser geodatabases served by ArcSDE.

When you expand a database connection that represents a multiuser geodatabase, it contains the same types of datasets and feature classes as a single-user geodatabase.

For many years, ArcInfo coverages have been used to represent vector data. The coverage format has enjoyed widespread implementation at governmental agencies, corporations, and organizations throughout the world because it efficiently stores spatial and topological data; attribute data is stored in relational tables that can be customized and joined with other databases.

Coverages combine spatial data and attribute data and store topological associations among features. Spatial data is held in binary files and attribute and topological data is kept in INFO™ tables. The catalog combines the representation of coverage binary files and INFO tables into coverage feature classes.

The introduction of geodatabases in ArcInfo supplements but does not replace coverages. Where they are employed, coverages still most often serve the intended requirements of applications. Coverages can be displayed, queried, analyzed, and edited in the new ArcInfo applications.

You can choose to migrate coverages into geodatabases when the benefits of integrating feature behavior and storing all data in a database outweigh the effort of conversion. You can think of a geodatabase as a next-generation coverage.

WORKSPACES AND GEOGRAPHIC DATA

ArcInfo workspaces contain the three basic representations of geographic data—coverages contain vector data, grids contain raster data, and TINs contain triangulations that represent surfaces. Most data stored in a workspace implements the georelational model where topology is stored and attributes are linked to features.

An ArcInfo workspace is a special type of folder where attributes for data are stored in INFO tables and all of the tables are managed through an INFO subfolder that is invisible in the catalog. When you use the catalog to create, move, and delete items in an ArcInfo workspace, their integrity is maintained for you. You should never use Windows Explorer or My Computer to manage coverages, grids, or TINs; the synchronicity between coverages and the INFO subfolder will be broken and data corrupted.

COVERAGES, FEATURES, AND TOPOLOGY

Coverages contain feature classes that are homogeneous collections of features.

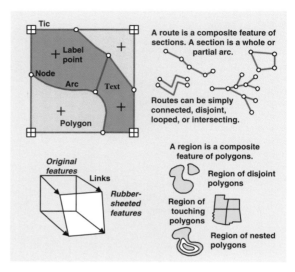

The primary types of coverage features are points, arcs (lines), polygons, and nodes. These features have topological associations: arcs form the perimeter of polygons, nodes form the endpoints of arcs, points mark the interiors of polygons. Point features have a dual identity; they can represent small geographic objects such as wells and buildings and they can mark polygon interiors.

Secondary types of coverage features are tics, links, and annotation. Tics are used for map registration, links are used for adjusting features, and annotation is used to label features on a map.

Coverages also contain composite features. Routes are collections of arcs with an associated measurement system. A common use of routes is for transportation systems. Regions are collections of polygons that can be adjacent, disjoint, or overlapping. Regions are used for land-use and environmental applications.

The catalog's view of an ArcInfo workspace

 A catalog.

 A folder connection.

 A folder with coverages, grids, and TINs is called an ArcInfo workspace. Attributes for most of these feature classes are stored in INFO tables.

 A coverage is an integrated set of feature classes that contain topology. This icon denotes a coverage with polygon topology.

 A point coverage contains a point feature class. It can optionally contain tic, link, and annotation feature classes.

 A coverage point feature class contains point features. Attributes are kept in point attribute tables (PAT).

 A coverage arc feature class contains line features that form a network or define polygon boundaries. Attributes are kept in arc attribute tables (AAT).

 A coverage route feature class contains composite line features with a linear measurement system. Attributes are kept in route subclass tables.

 A line coverage contains an arc feature class. It can optionally contain node, route, point, tic, link, and annotation feature classes.

 A coverage polygon feature class contains areal features formed by a ring of arcs and interior label points. Attributes are kept in polygon attribute tables (PAT).

 A coverage region feature class contains composite areal features formed by a number of polygons. Attributes are kept in region subclass tables.

 A coverage label feature class contains label points that mark polygons. Each polygon contains one label point.

 A coverage node feature class contains node features that occupy the ends of arcs. Attributes are kept in node attribute tables (NAT).

 A polygon coverage contains a polygon and label point feature class. It can optionally contain region, arc, node, route, tic, link, and annotation feature classes.

 A coverage annotation feature class contains text on a map. Attributes can optionally be kept in text subclass tables.

 A coverage link feature class contains vectors used for adjusting local areas to known reference points. This is called rubber sheeting.

 A coverage tic feature class contains points used for registering maps that are digitized.

 A TIN is composed of points with z-values organized into a mosaicked set of triangles to represent a surface.

A grid is a rectangular matrix of cells that represent imaged or sampled data. Attributes for values are kept in a value attribute table (VAT).

An INFO table is a relational database table. Feature attribute tables are INFO tables linked with features by object identifiers.

While topological datasets such as geodatabases and coverages provide a foundation for rich geographic analysis and map display, many map uses can be satisfied with a simpler form of feature data.

Simple feature classes store the shapes of features with points, lines, and polygons, but do not store topological associations. This structure has the advantage of simplicity and rapid display performance, but the disadvantage of not being able to enforce spatial constraints.

For example, if you are making a parcel map, you want to ensure that the polygons forming parcels do not overlap or have gaps between them. Simple feature classes cannot ensure this type of spatial integrity.

Yet, simple feature classes comprise a large part of available geographic data because they are easy to create and are sufficient for geographic data that forms background layers on maps.

The geodatabase can contain simple feature classes. ArcInfo also supports interaction with shapefiles and CAD drawings in AutoCAD® and MicroStation® format, common repositories of simple feature data.

SHAPEFILES

ArcView GIS 2, an ESRI software product for map display and query, introduced the shapefile format to satisfy the need for simple feature datasets.

A shapefile is composed of three main files that contain spatial and attribute data. A shapefile can optionally have other files with index information. In the catalog, all these files that comprise a shapefile appear as one feature class.

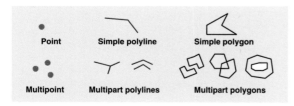

A shapefile is a homogeneous collection of features that can have either point, multipoint, polyline, or polygon shapes.

A point shapefile contains features with *point* geometry. A point is a single coordinate value.

A multipoint shapefile contains features with multipoint geometries, in which several points represent one feature.

A line shapefile contains features with *polyline* geometry. Polylines are made of *paths,* which are simply connected sets of line segments. The paths in a polyline can be connected, disjoint, or intersecting.

A polygon shapefile contains features with polygon geometry. A polygon contains one or many *rings.* A ring is a closed path that cannot intersect itself. The rings in a polygon can be disjoint, nested, or intersect one another.

While shapefiles store attributes in an embedded dBASE file, attributes of other objects can be stored in another dBASE table and can be joined to a shapefile by an attribute key.

CAD DRAWINGS

A substantial amount of geographic data has been collected in CAD (computer-aided design) drawing files. A characteristic of CAD files is that features are typically subdivided into many layers.

"Layer" in a CAD file has a different meaning than "layer" in a map. In a CAD file, it represents a set of similar features. In a map, it represents a reference to a geographic dataset or feature class with an associated drawing method.

A CAD dataset is the catalog's representation of CAD drawing files. It is subdivided into CAD feature classes, each of which aggregates all of the layers for points, lines, polygons, or annotation. If a CAD dataset has 17 layers—three with points, eight with lines, four with polygons, and two with annotation, they will be combined into a CAD point feature class, a CAD line feature class, a CAD polygon feature class, and a CAD annotation feature class.

ArcInfo supports interaction with CAD files in certain AutoCAD and MicroStation formats. Consult the online help for details on supported CAD formats.

The catalog's view of shapefiles and CAD drawings

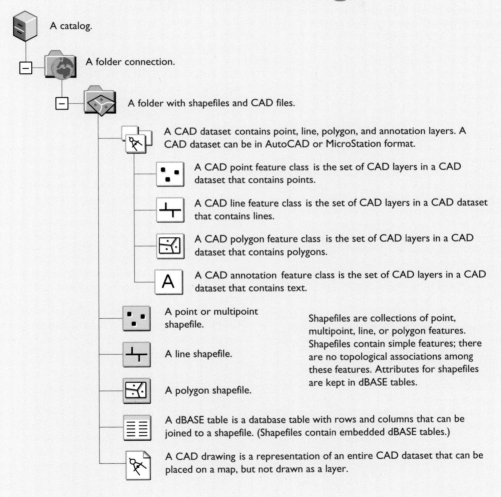

A catalog.

A folder connection.

A folder with shapefiles and CAD files.

A CAD dataset contains point, line, polygon, and annotation layers. A CAD dataset can be in AutoCAD or MicroStation format.

A CAD point feature class is the set of CAD layers in a CAD dataset that contains points.

A CAD line feature class is the set of CAD layers in a CAD dataset that contains lines.

A CAD polygon feature class is the set of CAD layers in a CAD dataset that contains polygons.

A CAD annotation feature class is the set of CAD layers in a CAD dataset that contains text.

A point or multipoint shapefile.

A line shapefile.

A polygon shapefile.

Shapefiles are collections of point, multipoint, line, or polygon features. Shapefiles contain simple features; there are no topological associations among these features. Attributes for shapefiles are kept in dBASE tables.

A dBASE table is a database table with rows and columns that can be joined to a shapefile. (Shapefiles contain embedded dBASE tables.)

A CAD drawing is a representation of an entire CAD dataset that can be placed on a map, but not drawn as a layer.

The catalog not only provides access to geographic data, it lets you manage the files that store your map and layer definitions. These files permanently store all the cartographic specifications you make in your maps.

Maps and layers in the catalog make it possible for you to create high-quality maps without writing any macro code. They also enable your organization to standardize the format, content, and appearance of your finished maps.

MAP DOCUMENTS, TEMPLATES, AND STYLES

Whenever you create a map in ArcMap, it is stored as a file on your computer disk with a .mxd file extension. This is called a *map document*.

A map stores the cartographic elements that comprise a map, but does not store geographic data. Instead, the layers in the map reference geographic datasets located at any location on a disk or database accessible by the catalog.

A *map document template* is a starting point for making any kind of map. It can be fairly simple with a set page size and style or it can be more complex with many cartographic elements and layers predefined. Templates make it easy for you to repetitively create a series of maps with a consistent appearance.

A *style* is a collection of palettes of cartographic objects that you use to draw maps. These objects include the marker symbols you use to draw point features, line symbols to draw line features, fill symbols to draw polygon features, and text symbols to draw annotation. Other objects in a style include colors and certain cartographic elements such as north arrows.

The purpose of styles is to ensure consistent use of cartographic symbols on your maps. Your organization can have multiple styles for generating different types of map products.

LAYERS

Layers can either be stored within a map document or in a separate layer file with a .lyr file extension. When you make simple maps, it will be most expedient for you to simply create your layers within the map, but when you want to share layers with other people, it is better to create them as separate layer files.

Since layers are references to geographic data, when a geographic dataset is moved or renamed in the catalog, simply updating the layer with the new location of the geographic data guarantees that all maps that include that layer still have the correct reference to data.

Map and layer files can be stored anywhere on your computer or network. They can be organized in a folder with geographic data or in a separate folder of their own.

The layers that you see in the catalog are only the layers that are stored in stand-alone files. To see the layers embedded in a map document, you will go to the table of contents for a map in the ArcMap application.

Vector, raster, and TIN layers

Point, line, and polygon layers can reference any feature class with zero-, one-, or two-dimensional feature geometries.

Point layers can reference points and junctions in a geodatabase; label points, tics, and nodes in a coverage; and points in a shapefile or CAD dataset.

Line layers can reference lines and edges in a geodatabase; arcs and routes in a coverage; and lines in a shapefile or CAD dataset.

Polygon layers can reference polygons in a geodatabase; polygons and regions in a coverage; and polygons in a shapefile or CAD dataset.

Annotation layers can reference text in a geodatabase or coverage.

Raster layers can be referenced by grids in an ArcInfo workspace, and image files in a variety of formats.

TIN layers can be referenced by TINs in an ArcInfo workspace.

The catalog's view of maps and layers

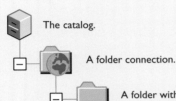
The catalog.

A folder connection.

A folder with maps and layers.

Layers can either be resident within a map document or can be a stand-alone file in any folder.

 A map document stores the cartographic elements that make up a map: title, north arrow, legend, scale bar, data frames, and layers.

 A map document template contains preset cartographic elements to implement mapping standards. Map documents are made from templates.

 A group layer is a collection of layers that are grouped for convenient placement on a map.

 A point layer references a feature class with single-coordinate features from geodatabases, coverages, shapefiles, CAD files, or other datasets.

 A line layer references a one-dimensional feature class such as a line or edge feature class in a geodatabase, a coverage arc feature class, or a line shapefile.

 A polygon layer references a two-dimensional feature class such as a polygon feature class in a geodatabase or coverage or a polygon shapefile.

 An annotation layer references an annotation feature class in a geodatabase, coverage, or CAD dataset.

 A TIN layer references a TIN dataset in a geodatabase or ArcInfo workspace.

 A raster layer references a raster in a geodatabase, a grid in an ArcInfo workspace, or an image file in a folder.

 A CAD layer references a CAD drawing in several common formats.

The three major types of geographic datasets that you work with in ArcInfo are geodatabases, coverages, and shapefiles. Because of how they are implemented in folders and databases and how topological information is stored, their data objects are structured differently.

This table summarizes the structure of vector data objects in geodatabases, coverages, and shapefiles and the functions they support. If you already use coverages and shapefiles, this table clarifies how they compare with geodatabases.

	Geodatabase	Coverage	Shapefile
Data collection	A geodatabase is a collection of feature datasets, rasters, and TINs.	An ArcInfo workspace is a collection of coverages, grids, and TINs.	A folder can contain shapefiles.
	All spatial, topological, and attribute data is stored in tables in a relational database.	Spatial data is stored in binary files. Topological and attribute data is stored in INFO tables.	Spatial data is stored in binary files. Attribute data is stored in dBASE tables. No topological data is stored.
	Geodatabases span continuous geographic extents.	For large datasets, coverages are subdivided into tiles in a map library.	Shapefiles are continuous for small to moderately sized datasets.
	Behavior is tightly coupled with features through rules and code written for custom feature classes.	Behavior is loosely coupled with features through AML scripts or VBA macros.	Behavior is loosely coupled with features through VBA macros.
Feature dataset	A feature dataset in a geodatabase contains simple or topological feature classes.	A coverage contains topological feature classes that participate in line or polygon topology.	A shapefile has one simple feature class.
	Line topology is implemented through a geometric network. Polygon topology is implemented through on-the-fly topological editing.	Line topology is implemented with arcs, nodes, and routes. Polygon topology is implemented with arcs, label points, polygons, and regions.	Polygon topology among a set of shapefiles can be implemented with on-the-fly topological editing.
	Many feature classes can be associated with a topological role.	Only one feature class is associated with a topological role.	There is no implicit topological role for a shapefile.
	User-defined associations can be established between features in different feature classes.	No associations are defined except for topological associations among related features like arcs and polygons.	No associations are established among features in shapefiles.
	Feature datasets have a defined coordinate system.	Coverages have a defined coordinate system.	Shapefiles have no defined coordinate system.
Feature class	A feature class stores features in a relational table with a special field for the geometric shape of a feature.	A coverage feature class stores feature geometry in a binary file and attributes and topology in a feature attribute table.	A shapefile stores feature geometries in a binary file and attributes in a dBASE file.
	The types of feature classes are point, line, polygon, annotation, simple junction, complex junction, simple edge, and complex edge.	The primary coverage feature classes are point, arc, polygon, and node. Secondary feature classes are tic, link, and annotation. Compound feature classes are region and route.	The types of shapefiles are multipoint, point, line, and polygon.
	A feature class can be extended to a custom feature class.	A coverage feature class cannot be extended.	A shapefile cannot be extended.

ArcInfo has a geometry model that represents the shapes of features in geodatabases, coverages, and shapefiles. The principal building blocks of this model are points; segments that can be straight, circular, elliptical, or Bézier curves; paths that are a set of connected segments; and rings that are a closed nonintersecting path. A polyline is composed of one or many paths. A polygon is composed of one or many rings.

Features in a geodatabase implement the full geometry model. Features in coverages and shapefiles implement a subset of the geometry model.

	Geodatabase	Coverage	Shapefile
Point features	+ Point Multipoint ++ + A feature class can contain features with point shapes or multipoint shapes. A multipoint is a set of points that represent one feature. Network junction features are also points.	+ Label points are point features or centroids of polygons with attributes. In a coverage with polygon topology, each polygon must have exactly one label point. Nodes are endpoints of arcs. They can have attributes. Tics are for registration. A coverage cannot contain multipoint features.	+ Point Multipoint ++ + A shapefile can have features that are simple points or multipoints. Points have no association with polygons.
Line features	In a geodatabase, a polyline has one or many paths. Paths are composed of four types of segments: line, circular arc, elliptical arc, Bézier curve A geometric network contains junctions and edges that form a one-dimensional network.	Arcs are simply connected sets of straight line segments with nodes at the endpoints. L R Arcs also participate in 2-D topology. They carry information about which polygons are to the right and left. Routes are composites of many sections. A section is a whole or partial arc. Routes have arbitrary connectivity.	Polyline with one path Polyline with several paths A shapefile has polylines that have one or many paths. There are no topological associations in a shapefile.
Polygon features	A polygon is made from one or many rings. A ring is a closed, nonintersecting path. Like polylines, polygons can have lines, circular arcs, elliptical arcs, and Bézier curves. Polygon with one ring Polygon with disjoint rings Polygon with nested rings	A polygon feature class is a planar graph with simple polygons. Each polygon has a label point, often at the centroid. Attributes are associated with label points. A planar graph is a continuous map of an area by nonintersecting polygons. Each point in an area is covered by exactly one polygon. A region subclass is a composite of polygon features.	Polygons in shapefiles are structurally similar to polygons in geodatabases except that the segments can only be straight lines. Polygon with one ring Polygon with disjoint rings Polygon with nested rings

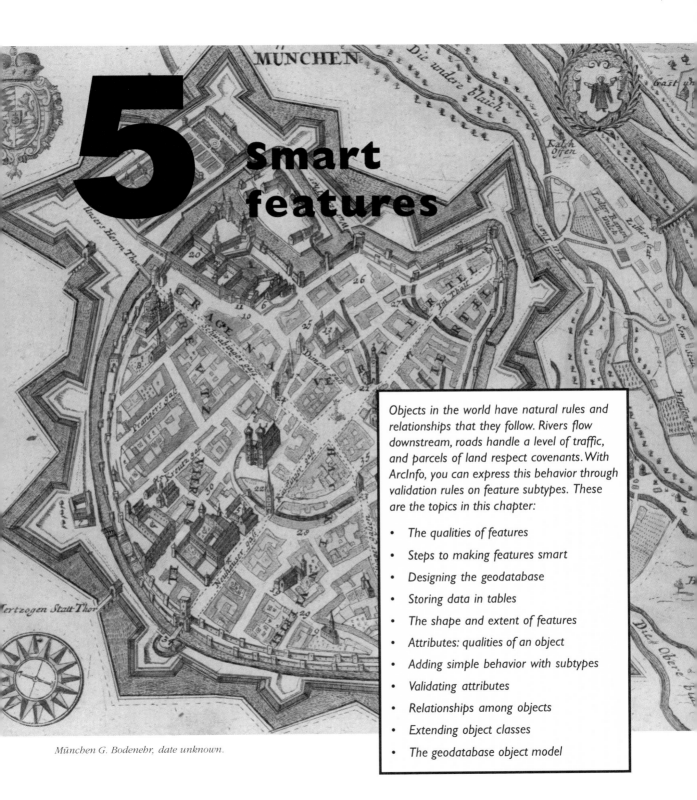

5 Smart features

Objects in the world have natural rules and relationships that they follow. Rivers flow downstream, roads handle a level of traffic, and parcels of land respect covenants. With ArcInfo, you can express this behavior through validation rules on feature subtypes. These are the topics in this chapter:

- The qualities of features
- Steps to making features smart
- Designing the geodatabase
- Storing data in tables
- The shape and extent of features
- Attributes: qualities of an object
- Adding simple behavior with subtypes
- Validating attributes
- Relationships among objects
- Extending object classes
- The geodatabase object model

München G. Bodenehr, date unknown.

Geographic objects in the world exist within a rich context. They occupy a position, delineation, or area; have neighboring objects; may influence surrounding objects as a consumer or provider of a resource; have attributes that can be values, counts, categories, or descriptions; and might have a predictable response to an external stimulus.

FEATURES IN THE GEODATABASE DATA MODEL

Features, as they are represented in the geodatabase data model, have many qualities such as vector shapes, relationships, attributes, and behaviors. These qualities collectively express the rich context that geographic objects experience.

For many applications, vector features are the most versatile geographic data representation, suited for geographic objects that have distinct boundaries and are persistent. Other geographic objects that can be considered continuous phenomena are better modeled with rasters or TINs.

This chapter discusses the qualities that make vector features smart in the geodatabase. First, a brief overview.

Features have shapes

The shape of a feature is stored as a special field in a feature class table of type *geometry*. A feature can be represented by one of these types of geometries:

- Points and multipoints, which are a set of points.
- Polylines, a set of line segments that may or may not be connected.
- Polygons, a set of rings that can be disjoint or embedded. A ring is a set of connected, closed, nonintersecting line segments.

The line segments that make up polylines and polygons can be straight line segments, circular arcs, Bézier curves, and elliptical arcs.

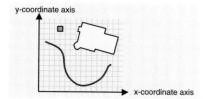

Features have a spatial reference

The shape of a feature is stored with x and y values in a Cartesian coordinate system. But the surface of the earth is roughly spherical. A spatial reference specifies how the x,y coordinates of a set of features are mapped onto the earth's surface.

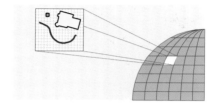

Features have attributes

A feature maintains its attributes as fields in a feature class table. Feature class tables are tables in a relational database. Attributes define standard and custom properties of features and can be numeric, textual, or descriptive.

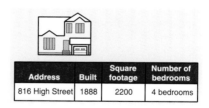

Address	Built	Square footage	Number of bedrooms
816 High Street	1888	2200	4 bedrooms

Features have subtypes

Features are collected into feature classes. Feature classes are homogeneous sets of features, but there may be considerable variation among features.

A feature class comprising buildings can be logically subdivided into subtypes such as residential, commercial, and industrial. Subtypes give you increased control of other qualities of features such as attribute domains and rules.

Features have relationships

All geographic objects have some relationship to other objects. You can define explicit relationships among geographic objects in different feature classes.

You can also define relationships to nonspatial objects, such as the relationship between a house and its owner.

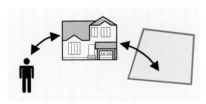

Feature attributes can be constrained

To enhance the accuracy of data collection, each attribute of a feature can have an attribute domain, which is a numeric range or a list of valid values. Each attribute can also have a default value automatically assigned when a feature is created. You can set distinct attribute domains and default values for each subtype in a feature class.

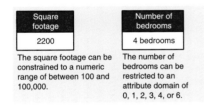

Square footage	Number of bedrooms
2200	4 bedrooms

The square footage can be constrained to a numeric range of between 100 and 100,000.

The number of bedrooms can be restricted to an attribute domain of 0, 1, 2, 3, 4, or 6.

Features can be validated by rules

Objects in the world follow rules when they are placed or changed. You can use rules to constrain how the parts of a network are connected or the cardinality of relationships.

A 6-inch pipe can be connected to a 4-inch pipe only with a proper fitting.

Relationships between houses and owners can be restricted to two owners per house.

Features can have topology

Many types of features have a precise relationship that is characterized as topology.

Parcels of land within a subdivision must adjoin each other exactly, without gaps or overlaps. This two-dimensional graph is called a *planar topology*.

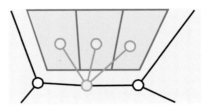

The lines and devices of a utility network must be continuously and unambiguously connected. This one-dimensional graph is called a *geometric network*.

Features can have complex behavior

Simple behaviors of features are implemented by choosing a feature type and topological association, setting up relationships, assigning attribute domains, and specifying validation rules.

More complex behaviors of features can be implemented by extending a standard feature and writing software code for a *custom feature*. Custom features permit complex behaviors such as custom editing interaction, intrinsic analytical capabilities, and sophisticated cartographic rendering.

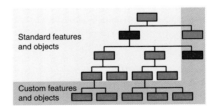

Standard features and objects

Custom features and objects

SMART FEATURES

Features in a geodatabase have a framework of attributes, geometry, spatial reference, relationships, domains, validation rules, topology, and custom objects. All aspects of this framework, except for complex behavior, require no programming.

Features in the geodatabase give you considerable control in modeling the world more naturally.

When you design and build a geodatabase, you employ a progressive set of steps to add intelligence to your features. As you have maximized the utility of each step, you move on to the next to further refine your feature model.

Depending on the requirements and complexity of your application, you may find that you may need to employ only a subset of techniques. For example, most applications will not require custom objects. And some applications may not require establishing relationships among features and objects; topological associations may do the job.

PROGRESSIVELY ADDING INTELLIGENCE

The following are the techniques for tailoring and customizing objects. The remainder of this chapter focuses on these techniques in more detail.

Select feature type and topology

Early in your implementation of a data model, you should take an inventory of all the types of objects you will need to model in your geodatabase. From this inventory, you will establish feature datasets to group feature classes that are bound by spatial reference, topology, and similar thematic content.

For nonspatial objects, create object classes. For spatial objects, create simple feature classes with point, line, or polygon shapes. For topological features, create a graph with topological feature classes in a common feature dataset.

Set attribution and subtypes

Once you've defined the type of object or feature class, you can add additional fields for the attributes of your object.

Objects and features can have a special attribute called a subtype. Subtypes are used for major groupings of objects and let you express diversity among similar objects or features without requiring that you create many object or feature classes.

A subtype for a type of road would let you model dirt roads, residential roads, and highways, and enforce data integrity specifically for each subtype. Subtypes improve your data integrity through attribute domains, default values, connectivity rules, and relationship rules.

Define attribute domains and validation rules

An attribute domain is a specified set or range of valid attributes. They prevent many simple mistakes when you apply a value to an attribute.

A default value applies an expected attribute value for a new object. It can streamline data entry by automatically assigning common attribute values.

Connectivity rules apply to features in a network; they are used to validate whether one type of feature can be correctly connected to another type of feature.

Establish object relationships

All objects interact with other objects. Important associations between objects that cannot be captured through topological associations can be captured as relationships. Relationships are stored in relationship classes and let you control and customize how objects and features are created, modified, and removed.

You can define relationship rules on relationship classes to further validate exactly how many features or objects can be associated with another.

Create custom objects

Object classes, domains, default values, validation rules, and relationships can express the majority of an object's desired behavior, but sometimes more complex behaviors for drawing, editing, or inspecting objects are needed. The set of ArcInfo object and feature classes can be extended by a programmer to create sophisticated and highly specialized objects and features.

SUMMARY

Most customization you will need for objects and features in ArcInfo can be done in the geodatabase data model without writing any software code.

As a data modeler, one of your main goals is to capture as much of the natural behavior of your objects as possible with this framework. Defining custom objects and features and writing software code should be necessary only for advanced applications.

The spectrum of tailoring features

simple

Select feature type and topology

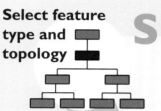

Determine the groupings of similar features and objects.

If there are topological associations, select topological feature types.

Otherwise, select simple feature types and specify geometry type.

Create object classes for nonspatial objects.

Set attribution and subtypes

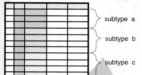

subtype a

subtype b

subtype c

Define the attributes for object and feature classes.

Decide whether an object class requires subtypes.

Assign names for subtypes.

Assign default values for attributes.

Set up attribute domains for valid values and numeric ranges.

Declare attribute update policies for splitting and merging.

Define relationship rules.

If features are in a network, define connectivity rules.

Define attribute domains and validation rules

Declare the types of relationships among object and feature classes.

Define optional attributes for each relationship type.

Constrain relationship cardinality with relationship rules.

Decide what should happen to a related object when a selected object is changed or removed.

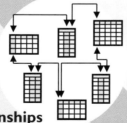

Establish object relationships

For complex behavior such as custom editing, complex validation, specialized drawing, or sophisticated analysis, extend the standard object or feature class types and write software code. Custom objects are necessary only for advanced data models and applications.

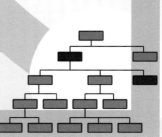

complex

Create custom objects

Because geographic features exist in a rich context with topology, spatial reference, and relationships, you have a number of decisions to make when designing your geodatabase.

These are the design considerations you should be aware of when you build your geodatabases.

CREATING GEODATABASES

You can work with any number of geodatabases in ArcInfo, but in certain situations grouping or splitting sets of features by geodatabase is better.

These are some reasons to group a set of features into a common geodatabase:

- If a set of objects and features have relationships, they must be in a common geodatabase.
- Features that have topological associations must be in a common geodatabase.
- If you need to concurrently edit a set of features, they must be in a common geodatabase. You can view multiple geodatabases in ArcMap, but you can edit only one geodatabase at a time.

These are some reasons to separate a set of features into distinct geodatabases:

- If you are working in a large organization, different departments have responsibility for various datasets. Geodatabases can be deployed to follow your organizational structure.
- You have the freedom to use any number of commercial relationship databases, but each must be served through a separate geodatabase.
- If you are working with personal geodatabases, practical size limits may require thematic or spatial partitioning of geodatabases.

ORGANIZING FEATURE DATASETS AND CLASSES

A geodatabase contains three general types of classes: object, feature, and relationship. These classes can either reside in a feature dataset or as stand-alone classes in a geodatabase. These are some reasons to group classes in a feature dataset:

- If feature classes are topologically related by a geometric network or planar topology, they must reside within a common feature dataset.

- If you want to enforce a common spatial reference for a set of feature classes, they should be in a common feature dataset.
- You can also arbitrarily group thematically related classes in a feature dataset.

There is no restriction on the placement of relationship classes; they can reside anywhere within a geodatabase and represent origin and destination classes throughout the geodatabase. If the origin and destination classes of a relationship class are in a common feature dataset, that is a good location for the relationship class, but it is not required.

APPLYING SUBTYPES

One of the most important design decisions you will make is whether a group of related features should constitute a set of feature classes or a single feature class with features segregated by subtype.

A subtype is a lightweight classification of features (or objects) within a feature (or object) class. The key motivation for using subtypes is performance. A geodatabase with one or two dozen feature classes will perform better than a geodatabase with many dozens of feature classes.

Subtypes let you control specific behavior for a set of features in a feature class through attribute rules, default values, connectivity rules, and relationship rules. Whenever possible, your first preference should be to use subtypes to differentiate groups of related features.

These are some reasons why it is sometimes necessary to split groups of related features into distinct feature classes:

- When each group of related features requires distinct custom behavior.
- When the set of feature attributes is different. (All features in a feature class have the same set of attributes.)
- When you require different access privileges for each group of features.
- When some features are to be accessed through versions and some are not.

Structure of features and objects

This conceptual illustration shows how features and objects are structured in a geodatabase. At the end of this chapter is a UML diagram of the geodatabase structure from a programmer's perspective.

A geodatabase contains feature datasets, stand-alone object classes, feature classes, relationship classes, and attribute domains.

A feature dataset contains object, feature, and relationship classes, a spatial reference, and geometric networks.

Object classes store nonspatial entities and have subtypes and attribute rules. (In the geodatabase, "object class" and "table" are synonyms.)

Feature classes store simple or topological features and have subtypes and attribute rules. A stand-alone feature class can have only simple features.

Relationship classes store relationships, may have attributes, and have associated relationship rules.

All feature classes in a feature dataset share the same spatial reference. Stand-alone feature classes also have a spatial reference.

All feature classes that participate in a geometric network must be in a common feature dataset. Each geometric network has a set of connectivity rules.

Any feature classes that you want to perform two-dimensional topological editing on must be in the same feature dataset. This ensemble of feature classes is called a planar topology.

Note: At the initial release of ArcInfo 8, planar topologies are not yet persistently stored in the geodatabase, but are dynamically defined when performing topological editing in ArcMap.

Attribute domains are stored directly in the geodatabase, ready to be applied to any attribute of any object or feature class in the geodatabase. When it is applied, it becomes an attribute rule for that object or feature class.

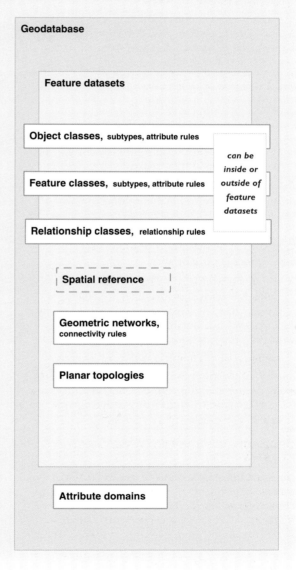

Tables are the repository of objects and their attributes. A table stores attributes for objects that are reasonably similar to one another and have the same set of attributes. For example, a table could store records for persons, buildings, and roads.

TABLES AND ROWS

A *table* is organized into rows and columns.

A *row* is the fundamental unit of information in a table and comprises a set of properties for an object. All the rows in a table must have the same set of property definitions.

A *column* represents all the attributes of one type. The value of a column for a given row is called an attribute. The definition of a column—its name and whether the column is formatted to store an object identifier, geometry, real numeric value, integer numeric value, or character string—is called a field.

Types of tables

In a geodatabase, tables can store nonspatial objects, spatial objects, and relationships.

A table that stores nonspatial objects is called an object class. It has a special field for subtypes.

A table that stores spatial objects is called a feature class. Simple feature classes have two predefined fields: a feature ID and a geometry field. Annotation feature classes and network feature classes have additional predefined fields.

A table that stores relationships is called a relationship class. It can have any number of custom fields to represent the attributes of the relationship. Not all relationship classes are implemented as tables. If a relationship class is not attributed and does not have a cardinality of many-to-many, it is stored as a set of foreign keys on the feature or object classes. Attributed relationships or many-to-many relationships are stored in tables.

Fields in a geodatabase

Attributes can express several qualities of an object. These are some common types of attributes:

- An attribute can designate a coded value for a classification.

- An attribute can be descriptive text that characterizes a feature or gives its name.

- An attribute can characterize a real numeric value that is measured or calculated, such as distance or flow.

- An attribute can represent a counted value, such as the number of associated parts.

- An attribute can specify a unique identifier that references a row in another table.

A table in a geodatabase can support these and other types of attributes with these field types: *float, double, short integer, long integer, text, date, object ID,* and *binary large object (BLOB)*.

Predefined and custom fields

There are two sets of fields in a table: *predefined fields* for uniquely identifying objects and storing feature shapes, and *custom fields* for defining additional attributes of features. Predefined fields and custom fields coexist in the same feature class table.

Predefined fields are managed by ArcInfo and should never be modified through another database application.

Custom fields implement the various types of attributes needed to realize the properties of your features. In the diagram, custom fields describe road type, surface, width, lanes, and name. You can add any number of custom fields.

Attribute and spatial indexes

You can create attribute indexes on fields to make query performance quicker. In ArcCatalog, you can create indexes on one or several attributes in a table and you can add and remove indexes at any time. Be careful with indexes—performance diminishes when you have defined an excessive number of indexes.

ArcInfo automatically creates spatial indexes on feature classes. It determines and applies an optimum grid size for you. To optimize certain feature classes, particularly when feature size varies considerably, you can define up to three grid sizes for best retrieval of spatial data.

Tables, objects, and attributes

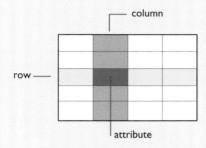

column

row —

attribute

A table is a set of rows.

A row is a set of attributes.

Columns represent all attributes of the same type in a table.

A field is a description of a column.

All rows in a table have the same set of fields.

An object class is a database table in a geodatabase. In addition to the basic functions on tables and their rows and columns, you can apply some of the functions of feature classes such as subtypes, attribute domains, default values, and relationships.

A feature class is an object class with a geometry field to keep the shape of features.

A feature class with line geometry has several predefined fields: a unique feature ID, a geometry-tracking field to record the feature length, and a geometry, which is the shape of the feature.

The other fields shown are custom fields. Some examples of custom attributes are coded values for road type, a descriptive string for surface type, a continuous value for road width, a discrete numeric value for the number of lanes, and text for names.

All attributes are based on one of these types: float, double, short integer, long integer, text, and date.

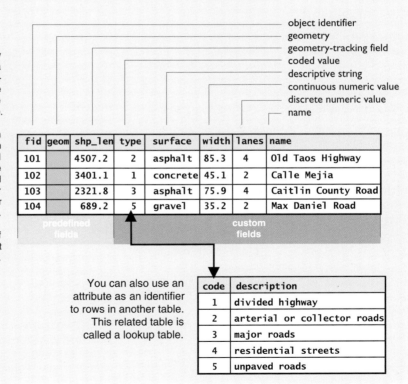

object identifier
geometry
geometry-tracking field
coded value
descriptive string
continuous numeric value
discrete numeric value
name

fid	geom	shp_len	type	surface	width	lanes	name
101		4507.2	2	asphalt	85.3	4	Old Taos Highway
102		3401.1	1	concrete	45.1	2	Calle Mejia
103		2321.8	3	asphalt	75.9	4	Caitlin County Road
104		689.2	5	gravel	35.2	2	Max Daniel Road

predefined fields

custom fields

You can also use an attribute as an identifier to rows in another table. This related table is called a lookup table.

code	description
1	divided highway
2	arterial or collector roads
3	major roads
4	residential streets
5	unpaved roads

You can think of a GIS as an extension of database technology that stores, manages, and updates spatial information. Features are spatial objects and much of the functionality inside ArcInfo involves the display, query, and editing of features.

A feature class has a special field that represents the shape and location of features. This field is called *shape* and is of the field type *geometry*. All features in a feature class have the same type of geometry.

FEATURES AND GEOMETRY

A shape field of a feature class can be one of the following types of geometry: point, multipoint, polyline, or polygon.

A feature with *point* shape has a single x,y or x,y,z coordinate value. A feature with *multipoint* shape has a number of x,y or x,y,z coordinate values. There is no order implied in the set of coordinates in a multipoint shape.

A feature with *polyline* shape has one or more *paths*. A path is a connected collection of segments, each of which can be one of these types of parametric curves: *line, circular arc, elliptical arc,* or *Bézier curve*. An optional z value (commonly an elevation) or m value (measurement distance) can be associated with a feature with polyline geometry.

A feature with a *polygon* shape has one or more rings. A *ring* is a connected, closed, and nonintersecting set of segments. Each segment can be of type line, circular arc, elliptical arc, or Bézier curve. A ring cannot intersect itself, but can intersect other rings in a polygon. Rings in a polygon can touch at any number of points. An optional z value (commonly an elevation) can be associated with a feature with polygon geometry.

An important change in ArcInfo 8 from previous releases is that single- and multiple-part geometries are now integrated in the same feature class. In the coverage data model, single-part lines were kept in arc feature classes, multipart lines in route feature classes, single-part polygons in polygon feature classes, and multipart polygons in region feature classes. Single-part and multipart geometries are no longer stored separately in the geodatabase.

Another important change is support for parametric segments: circular arcs, elliptical arcs, and Bézier curves. These types let you more accurately represent the feature shapes and are especially important for civil engineering applications.

A specific feature in a feature class can contain a *null geometry*. The data modeler may use null geometry to represent objects that are sometimes represented as explicit features and sometimes as implicit features within composite objects.

Chapter 6, "The shape of features," contains more information about the geometry of feature shapes.

FEATURES AND SPATIAL REFERENCE

The geometry of features is stored as a structured set of x and y coordinates, with optional z and m values. These coordinates are related to the shape of the earth through a spatial reference.

One part of the spatial reference is the coordinate system, which defines the mathematical projection of a planar area to the roughly spherical shape of the earth, called a *geoid*.

The other part of a spatial reference defines how a set of coordinates is related to its storage in a geodatabase as integer values. The geodatabase uses integers internally to prevent ambiguities when comparing locations and applying spatial operators. When you use ArcInfo, these integer values are converted to map units and you are not aware of them except when you need to define the spatial domain and scale of a spatial reference.

The spatial domain consists of minimum and maximum values for x and y, and optionally z and m. The scale defines how many integers correspond to a map unit. If the scale is 1,000, then the maximum precision is 1/1,000 of a map unit.

There is an inverse relationship between scale and spatial domain. If you select a very high value for the scale, the allowable spatial domain is restricted. As a rule of thumb, the product of the scale and the greatest range of the spatial domain cannot exceed two billion (or 2 to the 31st power).

For more information about spatial references, read the ESRI Press book *Understanding Map Projections*.

Feature geometry

Points	Lines	Polygons

Features in a geodatabase have one of four types of geometry: point, multipoint, polyline, and polygon.

Polylines are comprised of one or many paths. Polygons are comprised of one or many rings.

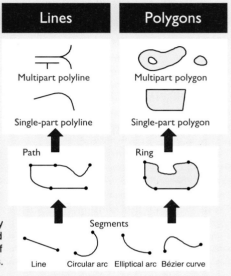

Multipoint

Point

Multipart polyline

Single-part polyline

Multipart polygon

Single-part polygon

A path is a simple, connected series of any of the four types of segments. A path cannot intersect itself.

A ring is a path that is closed.

Path

Ring

Four types of segments are presently supported in a geodatabase. Advanced developers can introduce new types of segments such as spirals.

Segments

Line   Circular arc   Elliptical arc   Bézier curve

Spatial reference

A spatial reference has three main components:

A coordinate system defines a map projection and its parameters.

A spatial domain constrains the range of x and y values, and optionally z and m values.

A scale defines how many integer units correspond to a mapping unit, and defines the coordinate precision.

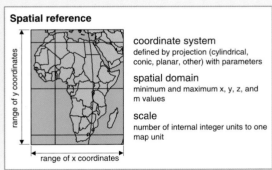

Spatial reference

range of y coordinates

range of x coordinates

coordinate system
defined by projection (cylindrical, conic, planar, other) with parameters

spatial domain
minimum and maximum x, y, z, and m values

scale
number of internal integer units to one map unit

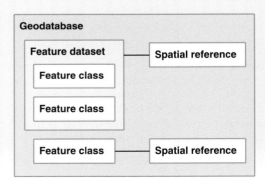

Geodatabase

Feature dataset — **Spatial reference**

Feature class

Feature class

Feature class — **Spatial reference**

A spatial reference is associated with a feature class. If feature classes are organized within a feature dataset, all of those feature classes share the same spatial reference.

A geodatabase can have many spatial references, one for each feature dataset and each stand-alone feature class.

Once a spatial reference is assigned to a feature class or feature dataset, you can update the coordinate system, but not the spatial domain or scale.

An attribute is a quality of an object. An attribute for a feature could be its size, density, name, flow, date of installation, or population.

Each object or feature in a GIS dataset has a number of attribute values, which are kept as rows in a database table. The attributes collectively represent the important qualities of that feature type for your application. Chapter 2, "How maps inform," discussed how attributes can be shown on a map. ArcMap supports a rich set of drawing methods to present attributes.

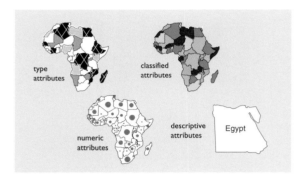

TYPES OF ATTRIBUTES

A number of types of attributes can be associated with a feature in a geodatabase. The following are some examples drawn from roads.

Continuous numeric values with floats and doubles

An attribute can contain a real numeric value, such as 2.7. These attributes are continuous data that is measured or calculated, such as distance or flow. A road has numeric attributes for its length, route measure, or width.

Discrete numeric values with shorts and integers

Some attributes represent numeric values that is counted, not continuous. They are usually positive integer values, but can be negative in some cases. A road has integer attributes for the number of lanes.

Coded values with shorts, integers, and text

An attribute can be a coded value. The value itself is not meaningful, but it references other attributes that are uniquely tied to the coded value. A road

feature may have an attribute designating road types in this fashion:

1—divided highway

2—arterial or collector roads

3—major roads

The advantage of coded values is that they take up little space in a table; their disadvantage is that you are one step removed from a meaningful attribute.

Descriptions with text

An attribute can be a descriptive string that characterizes a feature or gives its name. A road feature might have these allowable values for its surface type: "asphalt," "concrete," and "gravel."

A road feature would likely have an attribute for its name, such as "West Manhattan Avenue." This name might be kept in its entirety, or might be broken down into a road prefix ("West"), road name ("Manhattan"), and road suffix ("Avenue"). Exactly how a road name is represented is important if you are performing address matching in your GIS.

Time values with dates

Important events on objects can have a date value assigned that records a time. This is the temporal dimension of your GIS. When a road's pavement is replaced or maintained, a road inventory database can mark the time of that event with a date value.

Object identifiers

The power of a relational database is realized when relationships are made between rows or features from different tables. A feature attribute can be an identifier that references a feature or row in another table. For a road feature, an object identifier might reference an annotation with the road's name or a maintenance record of that road. When you create nonattributed relationships or annotation features, object identifiers are inserted for you.

Multimedia with BLOBs

Tables in a GIS can contain a BLOB (binary large object) column. This enables you to integrate other media such as video, images, or sound. A section of road can be associated with roadside photographs stored in a BLOB column in a table.

Attribute types

Applications

10.0
2.3
float

8.63
double

-4.7

A float value contains one sign bit, seven exponent bits, and 24 mantissa bits.

A double value contains one sign bit, seven exponent bits, and 56 mantissa bits.

Graduated symbols

Any type of numeric value can be drawn with graduated symbols, which vary in proportion to a value.

Classified values

A classification is a statistical subdividing of the numeric values of a set of objects. Classified values are drawn with color ramps.

-14
64
short integer
long integer

58
143

A short integer value contains one sign bit and 15 binary bits with a range of approximately −32 thousand to 32 thousand.

A long integer value contains one sign bit and 31 binary bits with a range of approximately −2 billion to 2 billion.

Unique values

Some attributes represent categories or types. A random set of symbols is applied to each unique value.

Sea
Blvd
Arkansas
Red
45th
text

| A | B | C | D | E | F | G | H | I | J | | | | | |

Text values contain any number of characters. Each character is stored in a byte (8 bits). All text values in a field have the same number of characters with trailing blanks.

Description

Egypt

Text shows names and other qualities of features.

12/1/61
1/30/
date
7/16/97

Date values are based on a standard time format.

The date value is translated into the current day and time in the local time zone.

635432
object ID
689764

An object ID value is a long unique identifier generated in geodatabases.

Object IDs are used for database joins and establishing relationships between objects.

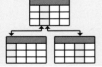

BLOB

BLOB values contain complex objects like images and video.

BLOB values let you add any kind of multimedia content to your geodatabase tables.

Every user's goal when adding or editing objects and features in a GIS database is to eliminate or minimize data entry error. For many modelers, this is the most important aspect of designing a geodatabase.

The easiest way to add intelligent behavior that verifies the integrity of features and objects is to:

- Apply constraints for updating attributes

- Define validation rules for how features are referenced or located with respect to each other

Further, you would like a fine level of control so that you can define behavior that discriminates between subgroups of feature classes, or subtypes.

Subtypes

Objects in an object class and features in a feature class may be further subdivided into subtypes.

A subtype is a special attribute that lets you assign distinct simple behavior for different classifications of your objects or features. All subtypes of a class share the same set of attributes.

The motivation for defining subtypes of an object class is to introduce a lightweight subdivision of an object class that adds these capabilities:

- You can name subtypes to describe each member of a classification of your objects.

- You can define distinct attribute domains for each field in a subtype.

- You can define distinct default values for each field in a subtype.

- You can prescribe the types of relationships that are possible between the objects in a subtype and objects in another subtype in the same or different object class.

- If you write some software code, you can also add custom rules for subtypes of object and feature classes.

An object class does not have to contain subtypes. If none are defined, you can still set attribute domains, default values, and rules—but on the object or feature class as a whole instead of on a subtype.

Attribute domains

Constraints on attributes are called attribute domains. For numeric attributes, you can set a range domain that constrains the value to between prescribed minimum and maximum values. An example is to constrain the price of one hectare of land to between 10,000 and 1,000,000 euros.

For all attribute types, except object IDs and BLOBs, you can set a coded value domain, which is a defined set of valid values. An example of coded value would be a list of geologic strata types such as "Pre-Cambrian," "Jurassic," and "Cretaceous." With a coded value domain, you can ensure that the attribute definitely has one of the expected values.

When editing features and objects in ArcMap, you can enter features with invalid values, but you can validate your work at any time. Invalid attribute values are highlighted for editing.

Validation rules

Validation rules control feature and attribute integrity. The types of validation rules are attribute rules, connectivity rules, and relationship rules.

An attribute rule is an attribute domain applied to a suptype of a class. An example of an attribute rule is that the field named DIAMETER can represent only pipes that are 10, 15, 25, or 50 centimeters in diameter.

A connectivity rule specifies the valid pairs of attribute values for subtypes for connected network features. For example, an electric line with phase ABC may be connected to a downstream line with phase AC. The types of connectivity rules are edge–junction rule, edge–edge rule, default junction type, and edge–junction cardinality.

A relationship rule constrains the cardinality of a relationship between an origin class and destination class. The four basic cardinalities are one-to-one, one-to-many, many-to-one, and many-to-many. With a relationship rule, you can create specialized cardinalities, such as a state has exactly two senators; a parcel of land can have no, one, or two owners; or a pole can have no, one, two, or three transformers mounted on it.

Simple behavior with subtypes

An object class or feature class can have a special type of field called a subtype. A subtype is used for what you regard as the most significant classification of the objects in your object class.

Subtypes help you preserve the integrity of your data.

fid	geom	subtype	width	lanes	name
101		asphalt	85.3	4	Chimayo Highway
102		concrete	45.1	2	Acequia de Isabel
103		asphalt	75.9	4	Calle Petra
104		gravel	35.2	2	Maximilian Road

For each subtype in your class, you can define simple object behavior with default values, attribute domains, connectivity rules, and relationship rules.

A subtype is stored as an integer value with a descriptive name such as "asphalt."

Features sorted by subtype

Roads with subtypes

fid	geom	subtype	width	ln	name
102		concrete	65	4	US Highway 285
103		concrete	75	4	NM Highway 14
104		concrete	75	4	US Interstate 25
101		asphalt	45	2	Grant Paige Ave
102		asphalt	35	2	Shakedown Street
103		asphalt	40	2	Acequia Wier
104		asphalt	45	2	Hart Alley
101		gravel	25	2	Garcia Road
102		gravel	15	1	Lesh Ranch Road
103		gravel	20	1	McKernan Lane
104		gravel	15	1	Kreutzman Road

Realizing simple behaviors

Validation rules

Default values	Attribute domains	Split/Merge policy	Connectivity rules	Relationship rules
A new concrete road is given a default value of four lanes.	Valid widths are 55, 65, and 75. Valid suffixes are "Highway" and "Interstate."	A split highway retains all highway designations.	A concrete road can connect to an asphalt road but not to a gravel road.	Two concrete roadways can be associated with a highway route.
A new asphalt road is given a default width of 35 feet.	Valid widths are 30, 35, 40, and 45. Valid lane counts are 1, 2, and 4.	A merged asphalt road takes a default value for lanes.	A two-lane asphalt road can only connect to another two-lane road.	An asphalt road can be related with bridges or tunnel crosses.
A new gravel road is given a default width of 15 feet.	Valid widths are 15, 20, and 25. Valid suffixes are "Road" and "Lane."	A split gravel road retains its width.	A gravel road cannot directly connect to a freeway.	A gravel road cannot have more than four road segments at an intersection.

Each subtype in an object class or feature class has a default value for new attributes, domains of valid attributes, rules to validate how features connect or relate, and the type of relationship possible for a new object.

An attribute domain is a constraint on attribute values in feature and object classes. This constraint can be a range of numeric values or a list of valid values.

Attribute domains are organized in a geodatabase and are ready for any object or feature class to use. When an attribute domain is applied to a particular attribute in a subtype, it becomes an attribute rule.

The following sections outline the components of attribute domains.

Range domains

To prevent data entry error, a range domain can constrain the value of any numeric attribute in any object or feature class to minimum and maximum values. An example of a range domain is the pressure in a pipe is expected to be between 2,000 and 14,000.

A range domain can be applied to short-integer, long-integer, float, double, and date attribute types.

Coded value domains

Many attributes are classifications of features. For example, a land-use type can be constrained to a list of values, such as "residential," "commercial," and "park." You can update the list of valid values in a coded value domain at any time.

A coded value domain can be applied to text, short-integer, long-integer, float, double, and date attribute types.

Default values

During data entry, it is frequently the case that for a certain attribute, one value is most commonly expected. Default values apply the expected value for a subtype in an object class when the feature is created, split, or merged. An example of a default value is applying "residential" as the default land-use classification of a new or split land parcel.

A default value can be applied to text, short-integer, long-integer, float, double, and date attribute types.

Splitting features

Once you have set a range or coded value domain, you can refine that domain by declaring what happens when features are split.

A land parcel split is a common scenario. When one piece of land is split into two, you might value the new parcels based on the proportion of their sizes. Or, you may want to apply an attribute value to both split parcels. You might also apply a default value to a new attribute.

The split policies are:

- Default value—A default value is applied to attributes of both split features.

- Duplicate—The attributes of both split features are identical to the value of the original feature attributes.

- Geometry ratio—You can define attributes of split features to be the proportional value of the split areas or lengths.

A split policy can be applied to text, short-integer, long-integer, float, double, and date attribute types.

Merging features

You can further refine an attribute domain by setting what happens to attributes when two objects are merged into one.

The merge policies are:

- Default value—A default value is applied to the merged features.

- Sum values—Two numeric attribute values are summed for the attribute of the merged feature.

- Weighted average—The attribute of the merged feature is the weighted average of the values of the attribute from the original features.

A merge policy can be applied to text, short-integer, long-integer, float, double, and date attribute types. An example of a merge policy is combining two crop yield values into one for a merged parcel of farmland.

Controlling attributes with domains

Attribute domains, default values, and split and merge policies are techniques that make it easy to validate attributes of features and objects.

Attribute domains

Range domain

pressure
2,000 to 14,000

A range domain specifies that an attribute value must be between a specified minimum and maximum value.

A range domain can be assigned to any type of attribute except text, object ID, or BLOB.

Coded value domain

land use
residential, commercial, park

A coded value domain is a set of valid values for an attribute.

A coded value domain can be assigned to any type of attribute except object ID or BLOB.

Default value

land zoning
R-1

A default value can be applied to an attribute whenever an object or feature is created.

A default value can also be applied when a feature is split or merged.

Splitting features

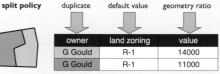

owner	land zoning	value
G Gould	R-4	25000

split policy   duplicate   default value   geometry ratio

owner	land zoning	value
G Gould	R-1	14000
G Gould	R-1	11000

attribute of original feature is duplicated in split features | default value is applied to split features | numeric attribute is subdivided by ratio of split area or length

Merging features

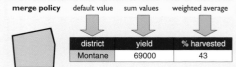

district	yield	% harvested
Lakeview	24000	35
Riverside	45000	47

merge policy   default value   sum values   weighted average

district	yield	% harvested
Montane	69000	43

default value for attribute is applied to the merged feature | numeric attribute is summed | numeric value is weighted average of attribute from the two features

Steps to setting attribute domains

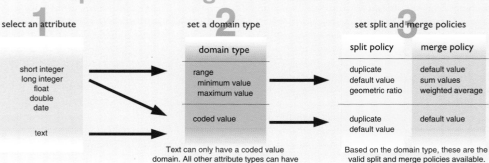

1 select an attribute

short integer
long integer
float
double
date

text

2 set a domain type

domain type
range minimum value maximum value
coded value

Text can only have a coded value domain. All other attribute types can have a coded value or range domain.

3 set split and merge policies

split policy	merge policy
duplicate default value geometric ratio	default value sum values weighted average
duplicate default value	default value

Based on the domain type, these are the valid split and merge policies available.

Objects in the world have relationships with other objects. Some objects have a spatial extent, such as roads. Other objects do not have a fixed spatial extent, such as people.

Some examples of relationships among objects are that a parcel of land can be associated with an owner, a land-use zone, an annotation with a lot number, or a building.

It is desirable to keep track of these relationships so that when one object is modified, the related objects can react appropriately. For example, when a utility pole is removed, the attached transformers and other equipment are removed as well.

The geodatabase provides a framework to explicitly define relationships among features and objects. ArcInfo includes the functionality to manage these relationships and ensure feature integrity.

WHEN TO USE RELATIONSHIPS

ArcInfo has three ways to define relationships among features: topological, spatial, and general.

What is connected or adjacent

Topological relationships are built into the data when you create a geometric network or planar topology. These relationships can quickly find neighboring polygons and traversed lines. They are managed for you through the topological environment of the ArcMap Editor.

What is spatially related

ArcInfo contains a rich set of spatial operations that can determine whether one feature touches, coincides with, overlaps, is inside of, or is outside of another feature. For example, you might want to determine which building footprints are inside of a land parcel.

General relationships

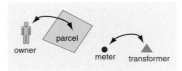

General relationships are explicitly defined relationships that form a persistent tie between a feature or object from an origin class to a feature or object in a destination class.

The data modeler can explicitly model general relationships between objects that cannot necessarily be inferred from their geometry or topology.

These are some uses of general relationships, again illustrated through examples of a road feature:

- A one-to-one relationship might be between a road feature and an associated row in an external table, such as road maintenance data.

- A one-to-many relationship might be between a road feature and a set of events, such as traffic accidents.

- A many-to-many relationship might be between many road features and many highway construction work orders.

Any of these relationships can be established between the features and objects in a geodatabase.

RELATIONSHIPS AND RELATIONSHIP CLASSES

A *relationship* is an association between two objects. These objects can be nonspatial (objects) or spatial (features). Besides identifying the associated objects, relationships can have additional properties.

Relationships are organized into *relationship classes*. Each relationship in a relationship class has the same origin class and destination class. Any object class may participate in many relationship classes. With relationships, the geodatabase ensures referential integrity between objects as they are created, modified, or deleted.

Relationships

The geodatabase uses relationships between objects to maintain the referential integrity of objects when they are deleted or moved. Related objects can issue notifications to each other when there is a change.

Pole–transformer relationship example

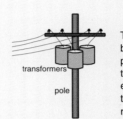

The association between utility poles and electric transformers is an example of a one-to-many relationship.

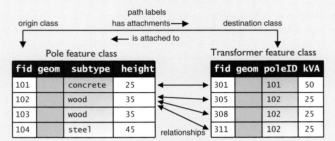

The pole feature class is considered the origin class because transformers are mounted on poles.

A pole–transformer relationship is considered a composite relationship because the placement and lifespan of a transformer is affected by the lifetime of the pole.

One pole can be mounted with zero, one, two, or three transformers, so a relationship rule would be defined to enforce this cardinality constraint.

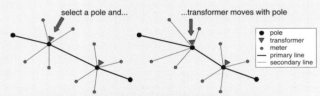

A composite relationship ensures that when an object from the origin class is moved or deleted, the related object from the destination class is also moved or deleted.

Cardinality of relationships

There are four basic cardinalities of relationships. The cardinality affects whether a relationship can be simple or composite.

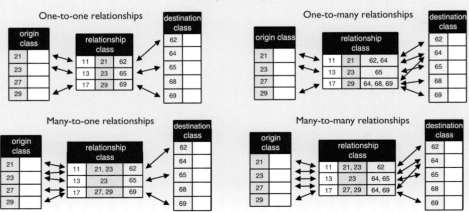

Path labels

A relationship class has a *forward path label* and a *backward path label*. These labels describe the relationship and are used when displaying object relationships. Examples of path labels are "manages" and "managed by."

Cardinality

A relationship class has *cardinality*, which restricts the number of relationships that can be formed between an origin object class and destination object class. Examples of cardinalities are one-to-one, one-to-many, many-to-one, and many-to-many.

A relationship class can be simple or composite.

A *simple relationship* is a peer-to-peer relationship where related objects can exist independently.

A *composite relationship* is a one-to-many relationship between a *composite object* from an origin class and *part objects* from a destination class. A composite relationship must have one-to-many cardinality, and you can use relationship rules to enforce this.

When an object from an origin class is deleted, the related objects in the destination class must also be deleted.

Part objects can be created independently of the composite object, but they must be deleted when the composite object is deleted. Part objects of a composite object can be deleted and replaced by new composite objects.

Notifications

A notification is a message passed in ArcInfo when a significant event occurs, such as an edit or deletion. Notifications are the mechanism that manages the lifespans of part objects based on the whole object in a composite relationship.

A relationship class may be used to propagate standard *notifications* between related objects. The *notification direction* property specifies these four notification options:

- No notifications are propagated.

- A notification is issued to the destination object only when the origin object is changed.

- A notification is issued to the origin object only when the destination object is changed.

- A notification is issued when either the origin or destination object is changed.

Attributed relationship classes

When ArcInfo creates a one-to-one or one-to-many relationship class, it implements it as inserted foreign keys on the origin and destination class. A relationship table is not built with these cardinalities unless you specify that you want to add attributes to relationships.

When ArcInfo creates a many-to-many relationship class, a relationship table must be built because the foreign keys cannot unambiguously record all of the relationships.

A relationship table has a row for each relationship. You can optionally add attributes to relationships. These attributes can be any qualities that describe an event that binds the related objects, such as a work order or accident report.

An example of an attributed relationship is between an owner and a piece of land. This relationship can represent a transaction in land ownership and the attributes could recite the facts recorded in the title document.

Cardinality	Control	Implementation
one-to-one	simple	managed with foreign keys or relationship table if relationships are attributed
one-to-many	simple or composite	managed with foreign keys or relationship table if relationships are attributed
many-to-many	simple or composite	relationship table is necessary

ANNOTATION AND ANNOTATION CLASSES

An annotation is a type of feature that provides a textual description of a place or feature.

An *annotation feature class* is a feature class that contains annotation. All annotation in an annotation feature class has the same set of attributes.

Feature-linked annotation

Most features on a map have annotation. Annotation is usually a place name, but can also be any attribute of the feature.

Annotation can be closely bound to features by defining a composite relationship between a feature class and annotation class. This is called feature-linked annotation.

The annotation feature class includes properties such as the field from which text labels are derived, the type of symbol, and other attributes.

When a feature in the composite feature class is created, a notification is sent to the annotation feature class, enabling automated placement of annotation. When a composite feature is deleted, the associated annotation is deleted. When changes are made to the composite feature, notifications are forwarded to the annotation class using the standard complex relationship notifications.

Simple annotation

Maps also have annotation that is not linked to a feature. Simple annotation can be used for:

- Map graticule information such as coordinate or latitude values

- Large or indeterminate geographic entities that are not represented by a single feature

- Any free-form text labeling on the map

Annotation

There are two types of annotation. Feature-linked annotation provides dynamic labeling of features. Simple annotation allows free-form placement of text on the map.

Feature-linked annotation

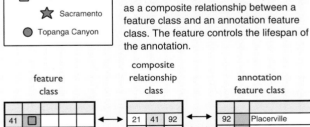

▣	Placerville
★	Sacramento
●	Topanga Canyon

Feature-linked annotation is implemented as a composite relationship between a feature class and an annotation feature class. The feature controls the lifespan of the annotation.

feature class → composite relationship class → annotation feature class

When a feature with feature-linked annotation is removed, the annotation is removed as well. When an attribute is changed, the text displayed by the annotation feature class is updated.

Simple annotation

Sierra Nevada Range	
Galisteo Basin	
Death Valley	

Simple annotation is kept in an annotation feature class that is not bound in a relationship with features.

annotation feature class

411	Sierra Nevada Range
413	Galisteo Basin
417	Death Valley

Simple annotation has no relationship to attributes of any other feature.

The highest form of customization in ArcInfo is creating custom features. Certain complex behaviors cannot be expressed with rules. These include custom editing, complex validation, specialized drawing, and sophisticated analysis.

If you are not a programmer, you can skip the remainder of the chapter. These pages discuss the concepts of programming complex behavior into custom features and the programmer's view of the geodatabase data access objects.

Object fundamentals

An *object* represents an entity such as a house, lake, or customer. An object is stored as a row and has behavior expressed as one or more sets of *methods*.

A method has a *name,* a set of *input parameters* and *output parameters,* and a *return type*. A set of methods is called a *software interface* (or simply, interface).

The developer of an object writes software code that provides implementation for methods. The ArcInfo system invokes these methods to express the complex behaviors of objects.

An *object class* represents a set of objects that share the same type, such as house, lake, or customer. The behavior of an object class is implemented with a *behavior class* that is stored in a dynamic-linked library (DLL) in conformance with the Microsoft Component Object Model (COM) architecture.

The ArcInfo object classes

ArcInfo provides a hierarchy of object classes ready for use. These are *object, feature, simple junction feature, complex junction feature, simple edge feature,* and *complex edge feature*.

The ArcInfo object classes include a number of predefined fields such as object identifiers and geometries. These fields define the required properties of the objects in these classes. The data modeler can add additional custom fields.

Each of these object classes implements a set of interfaces. Each interface contains a set of related methods for actions such as storing, editing, drawing, querying, and validating objects.

An ArcInfo object class provides default implementation for each of its interfaces. The developer can selectively override the default implementation for an interface.

Custom features

The developer can customize object classes by extending the ArcInfo object classes. ArcInfo includes a CASE (computer-aided software engineering) tool framework to graphically extend the standard ArcInfo object classes. This framework automates the schema generation for the new object classes and generates source code for behavior class templates.

The developer implements new interfaces or uses existing interfaces to model specialized behavior. The developer can also override existing software interfaces inherited from standard ArcInfo classes.

Type inheritance

Features can be specialized. You can create a new type of custom feature that has all the attributes and behavior of another, but adds new attributes and behaviors. For example, a state highway is a type of road. This is called *type inheritance.*

Features are COM objects. COM is an infrastructure to build software components on an inheritance model based on *interfaces.* An interface is a contract between server and client to provide specified services, such as drawing or selection.

In the illustration to the right, the Pole custom feature implements four interfaces: IPole, IFeatureDraw, IFeature, and IRow.

Internals of a custom object

A custom object is a combination of a database table and code compiled to a DLL.

A custom feature is implemented internally in ArcInfo as a feature class table, as a behavior class stored in a DLL, and as a globally unique identifier in the Windows Registry that binds the feature class and its behavior class.

Extending the object model

You can apply specialized behaviors of objects and features by writing software code that conforms to the interfaces and conventions of ArcObjects, the components inside ArcInfo.

Standard ArcInfo objects

These objects are part of the geodatabase data access objects and are supplied with ArcInfo. Most data modelers can build rich data models with only the standard feature and object types.

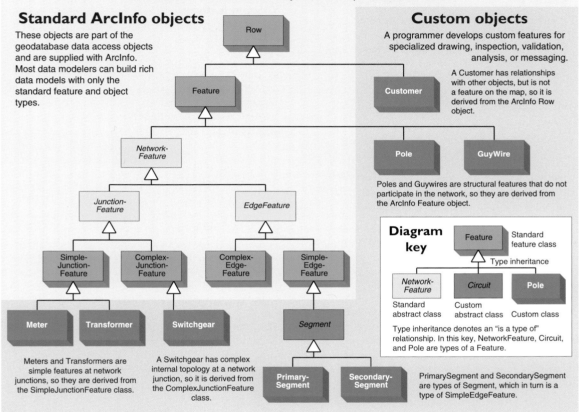

Meters and Transformers are simple features at network junctions, so they are derived from the SimpleJunctionFeature class.

A Switchgear has complex internal topology at a network junction, so it is derived from the ComplexJunctionFeature class.

Custom objects

A programmer develops custom features for specialized drawing, inspection, validation, analysis, or messaging.

A Customer has relationships with other objects, but is not a feature on the map, so it is derived from the ArcInfo Row object.

Poles and Guywires are structural features that do not participate in the network, so they are derived from the ArcInfo Feature object.

Diagram key

Feature — Standard feature class

Type inheritance

Network-Feature — Standard abstract class

Circuit — Custom abstract class

Pole — Custom class

Type inheritance denotes an "is a type of" relationship. In this key, NetworkFeature, Circuit, and Pole are types of a Feature.

PrimarySegment and SecondarySegment are types of Segment, which in turn is a type of SimpleEdgeFeature.

Type inheritance

A custom object is created by applying type inheritance using COM interfaces.

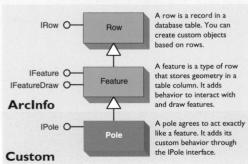

ArcInfo

IRow — Row: A row is a record in a database table. You can create custom objects based on rows.

IFeature, IFeatureDraw — Feature: A feature is a type of row that stores geometry in a table column. It adds behavior to interact with and draw features.

IPole — Pole: A pole agrees to act exactly like a feature. It adds its custom behavior through the IPole interface.

Custom

Internal implementation

The behavior for custom objects is implemented with COM interfaces in DLLs. The class table and DLL are linked through the Registry.

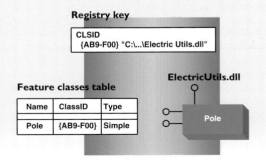

Registry key

CLSID
{AB9-F00} "C:\...\Electric Utils.dll"

Feature classes table

Name	ClassID	Type
Pole	{AB9-F00}	Simple

ElectricUtils.dll

Pole

Geodatabase data

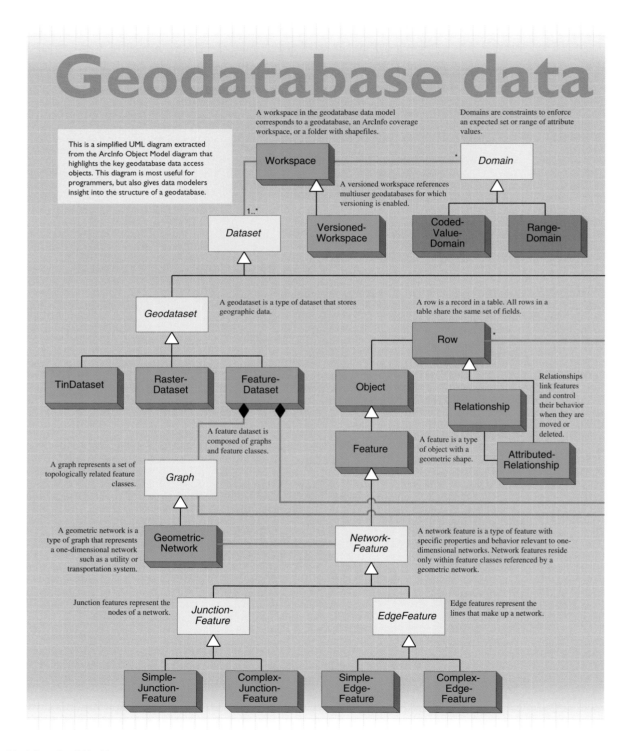

A workspace in the geodatabase data model corresponds to a geodatabase, an ArcInfo coverage workspace, or a folder with shapefiles.

Domains are constraints to enforce an expected set or range of attribute values.

This is a simplified UML diagram extracted from the ArcInfo Object Model diagram that highlights the key geodatabase data access objects. This diagram is most useful for programmers, but also gives data modelers insight into the structure of a geodatabase.

Workspace

A versioned workspace references multiuser geodatabases for which versioning is enabled.

Domain

1..*

Dataset

Versioned-Workspace

Coded-Value-Domain

Range-Domain

A geodataset is a type of dataset that stores geographic data.

A row is a record in a table. All rows in a table share the same set of fields.

Geodataset

Row

*

TinDataset

Raster-Dataset

Feature-Dataset

Object

Relationships link features and control their behavior when they are moved or deleted.

Relationship

A feature dataset is composed of graphs and feature classes.

A feature is a type of object with a geometric shape.

Feature

Attributed-Relationship

A graph represents a set of topologically related feature classes.

Graph

A geometric network is a type of graph that represents a one-dimensional network such as a utility or transportation system.

Geometric-Network

Network-Feature

A network feature is a type of feature with specific properties and behavior relevant to one-dimensional networks. Network features reside only within feature classes referenced by a geometric network.

Junction features represent the nodes of a network.

Junction-Feature

EdgeFeature

Edge features represent the lines that make up a network.

Simple-Junction-Feature

Complex-Junction-Feature

Simple-Edge-Feature

Complex-Edge-Feature

access objects

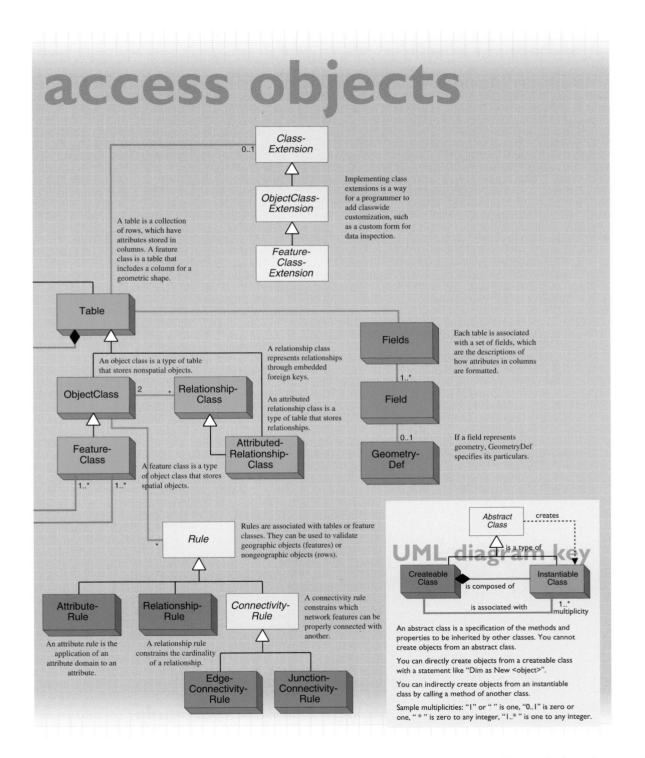

Class-Extension (0..1)

ObjectClass-Extension

Feature-Class-Extension

Implementing class extensions is a way for a programmer to add classwide customization, such as a custom form for data inspection.

A table is a collection of rows, which have attributes stored in columns. A feature class is a table that includes a column for a geometric shape.

Table

Fields

Each table is associated with a set of fields, which are the descriptions of how attributes in columns are formatted.

An object class is a type of table that stores nonspatial objects.

ObjectClass 2 * **Relationship-Class**

A relationship class represents relationships through embedded foreign keys.

An attributed relationship class is a type of table that stores relationships.

Field 1..*

Feature-Class

Attributed-Relationship-Class

A feature class is a type of object class that stores spatial objects.

1..* 1..*

Geometry-Def 0..1

If a field represents geometry, GeometryDef specifies its particulars.

Rule *

Rules are associated with tables or feature classes. They can be used to validate geographic objects (features) or nongeographic objects (rows).

Attribute-Rule

Relationship-Rule

Connectivity-Rule

A connectivity rule constrains which network features can be properly connected with another.

An attribute rule is the application of an attribute domain to an attribute.

A relationship rule constrains the cardinality of a relationship.

Edge-Connectivity-Rule

Junction-Connectivity-Rule

UML diagram key

Abstract Class — creates

— is a type of

Createable Class — is composed of — **Instantiable Class**

— is associated with — 1..* multiplicity

An abstract class is a specification of the methods and properties to be inherited by other classes. You cannot create objects from an abstract class.

You can directly create objects from a createable class with a statement like "Dim as New <object>".

You can indirectly create objects from an instantiable class by calling a method of another class.

Sample multiplicities: "1" or " " is one, "0..1" is zero or one, " * " is zero to any integer, "1..* " is one to any integer.

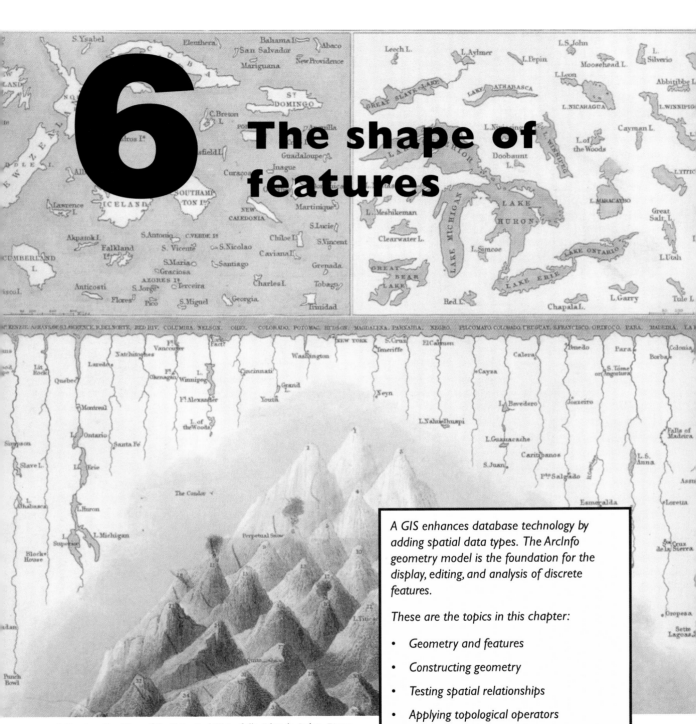

6 The shape of features

A GIS enhances database technology by adding spatial data types. The ArcInfo geometry model is the foundation for the display, editing, and analysis of discrete features.

These are the topics in this chapter:

- Geometry and features
- Constructing geometry
- Testing spatial relationships
- Applying topological operators
- Geometry object model

A Comparative View of the Principal Waterfalls, Islands, Lakes, Rivers, and Mountains in the Western Hemisphere, John Rapkin, 1851.

One of the primary data representation models of geography is the vector data model. In a geodatabase, vector data is implemented as features stored in feature datasets and feature classes.

Features offer several important advantages for data modeling:

- Features are stored as distinct entities with attributes, relationships, and behavior. This lets you create a rich model that captures all that you know about a set of geographic features.

- Features have precise locations with well-defined geometric shapes. In ArcMap, you can apply spatial operators that test for inclusion, overlap, or adjacency among a selection of features.

- Features can be drawn on a map with any color, line width, fill pattern, or other cartographic symbol. You can create maps that display feature attributes with symbols. You can also print a map of features with crisp detail at a wide range of map scales.

Features are especially well suited for modeling man-made objects. That is because roads, buildings, airports, and other built objects have sharp, well-defined boundaries.

The foundation of representing a feature in a geodatabase is its geometry, or shape. This chapter surveys the fundamentals of the ArcInfo geometry system and its key functions.

This chapter gives you insight in three ways:

- When you build your data model, you will understand how to best represent the shape of your features.

- When you edit maps in the ArcMap Editor, the section on geometry construction will help make you an expert map editor.

- If you are an application developer and need to customize how feature shapes are created and updated, this chapter gives you an overview of the geometry object model. This is not complete documentation of the geometry system, but it is an overview of the important concepts and software interfaces.

THE GEOMETRY SYSTEM

Each feature has a geometry (or shape) associated with it. In the geodatabase, the geometry is stored as a special field in a feature class that is called "shape."

In the geometry object model, there are two levels of geometries—those that define the shape of features, and those that are components of those shapes.

FEATURE GEOMETRIES

A feature can be created with one of these types of geometries: *point, multipoint, polyline,* and *polygon.* An *envelope* is a geometry that describes the spatial range of feature geometries.

An important advantage of the geodatabase data model over the coverage data model is that single-part and multipart geometries are combined within the same feature class. A feature class with polyline geometry can contain single-part or multipart polylines. A feature class with polygon geometry can contain single-part or multipart polygons. This gives you more freedom to model the shape of features and simplifies the structure of your geodatabase.

Points and multipoints

Points are zero-dimensional geometries. They have an x,y coordinate, with an optional altitude (z), measure (m), and point IDs. Points are used to represent small features such as wells and survey points.

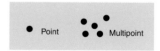

Multipoints are unordered collections of points. Multipoint features represent a set of points with a common set of attributes, such as a set of wells that form a single unit.

Polylines

A polyline is an ordered collection of paths that can be disjoint or connected. Polylines are used to represent the geometry of all linear features.

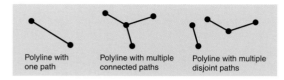

Polyline with one path | Polyline with multiple connected paths | Polyline with multiple disjoint paths

Polylines are used for roads, rivers, and contours. Simple linear features are represented by polylines with one path. Complex linear features, such as routes, are represented by polylines with many paths.

Polygon

A polygon is a collection of rings that are partially ordered by their containment relationship.

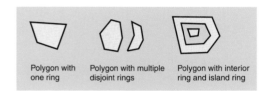

Polygon with one ring | Polygon with multiple disjoint rings | Polygon with interior ring and island ring

Polygons are used to represent the geometry of all areal features. Simple areal features are represented by polygons with one ring.

When rings are nested, they alternate between interior rings and island rings. Rings in a polygon can be disjoint but they cannot overlap.

Envelope

An envelope represents the spatial extent of features.

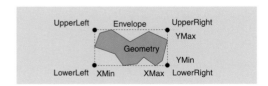

An envelope is a rectangle that spans the minimum and maximum coordinates of a geometry. An envelope also records the range of z and m values for a geometry. The sides of an envelope are parallel to a coordinate system.

All geometries have envelopes. Envelopes are used in ArcInfo to enable rapid display and spatial selection of features.

COMPONENTS OF FEATURE GEOMETRIES

Segments, paths, and rings are the geometries that are components of the feature shapes.

Segments

A segment consists of a start and endpoint and a function defining a curve between the points.

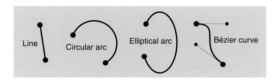

Line | Circular arc | Elliptical arc | Bézier curve

The four types of segments are lines, circular arcs, elliptical arcs, and Bézier curves.

- A *line* is a straight segment bounded by two endpoints. It is the simplest type of segment. Lines are used for straight constructions, such as a highway segment, or subdivisions of land, such as a parcel line.

- A *circular arc* is a portion of a circle. The most common use of circular arcs is for road curbs at street intersections. Circular arcs are widely used in COGO (coordinate geometry) applications. When a circular arc is part of a feature, it is nearly always tangent to the connecting segments.

- An *elliptical arc* is a portion of an ellipse. It is not frequently used for features, but can approximate transitional figures such as sections of a highway ramp.

- A *Bézier curve* is defined by four control points. It is a parametric curve defined by a set of third-order polynomials and is useful for depicting smoothly varying features such as contours and streams. Bézier curves are also used for the placement of text characters of a name along a meandering object such as a stream.

Paths

A path is a sequence of connected segments. The segments in a path cannot intersect. A path can contain any combination of lines, circular arcs, elliptical arcs, and Bézier curves. Paths make up polylines.

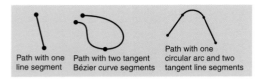

Path with one line segment | Path with two tangent Bézier curve segments | Path with one circular arc and two tangent line segments

Often, the segments that comprise a path are tangent to each other. That means that the segments join at the same angle. For example, a road is typically comprised of straight lines and circular arcs. When a line and a circular arc in a road join, they are at the same angle, or tangent to each other.

Another example of a path with tangent segments is an elevation contour, which is usually a set of connected and tangent Bézier curves.

Rings

A ring is a path that is closed and has an unambiguous inside and outside.

The coordinates for the start and endpoints of the path are the same. Rings make up polygons.

ATTRIBUTES OF FEATURE GEOMETRIES

When you create a feature class, you can assign up to three optional attributes to the vertices of feature geometries: a *z value,* an *m value,* and a *point ID.*

Vertical measurements with z values

The ArcInfo geometry system is fundamentally a two-dimensional system, but you can assign a z value for each point in a point, multipoint, polyline, or polygon.

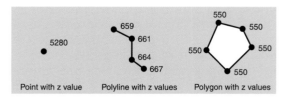

Point with z value | Polyline with z values | Polygon with z values

Z values most commonly represent elevations, but they can also represent another quality, such as rainfall level.

Z values can be applied to features such as stream lines, ridge lines, or lakes. A ridge line is a profile along a surface. You can assign individual elevations at each point along a ridge line. A lake polygon would have identical z values along the perimeter of the lake.

One use of z values is to prepare elevation data for input into a triangulated irregular network (TIN). Another use of z values could be a civil engineering application that models the vertical profile of a road alignment.

Linear measurements with m values

Some applications employ a linear measurement system that is based on interpolated distances along paths. You can assign an m value, or linear measure, to each point in a point, multipoint, polyline, or polygon.

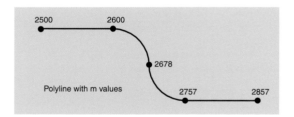

Polyline with m values

An example of a linear measurement system is mileposting, or stationing, along a road or canal. The geometry system has functions to interpolate m values for x,y points along a path or to calculate x,y positions from an m value along a path.

Managing points with point IDs

Sometimes, unique identifiers are collected with points. For example, each point collected with a survey instrument often has a point number. You can use point IDs to preserve any type of identifier that has been collected, and point IDs can be used in a custom ArcInfo application.

Note: At the initial release of ArcInfo 8, the ArcMap Editor does not directly edit z values, m values, or point IDs. When you edit features, the z values, m values, and point IDs that exist for a feature geometry are preserved. If a feature with m values is split, the m values of the points of the split features are interpolated correctly.

One of the important services of the ArcInfo geometry system is a rich set of construction methods that create new geometries from distances, angles, and relationships to existing geometries.

Many of these methods correspond to basic coordinate geometry (COGO) commands, which are useful for accurately converting survey data to geographic data. Other methods facilitate the editing of features to improve cartographic presentation or to control spatial relationships between features.

ArcInfo users will see most of these construction methods implemented as tasks within the ArcMap Editor. This section will give you insight into how the edit sketch in the ArcMap Editor works with the geometry system.

Programmers can build custom applications that modify the shapes of features by using the construction methods in the geometry object model. This section is not a complete reference to the geometry construction methods, but does list some of the most commonly used constructors.

Units and input

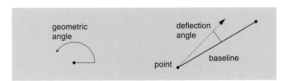

Angles are normally specified as geometric angles measured counterclockwise from the positive x axis of the Cartesian coordinate system. Some constructors use deflection angles that are measured from a point relative to a baseline.

If you are a programmer, all angles are specified in radians. If you are an ArcInfo user, angles are specified in degrees by default.

All distances are measured in map projection units.

Some constructors are built upon segments, paths, and rings. Recall their definitions: a segment is a line, circular arc, elliptical arc, or Bézier curve; a path is a sequence of connected segments; and a ring is a closed path.

Point construction

The methods that follow create a single point by specifying angles, distances, and relationships to existing geometries.

Programmers can access point constructors through the IConstructPoint interface implemented in the Point class.

Construct Along

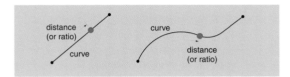

Given a curve and a distance or ratio, a point is constructed along that curve. If the distance is greater than the length of the curve, the point follows either a tangent or the embedding geometry.

Construct Angle Bisector

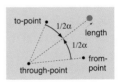

Given a from-point, through-point, and to-point, this constructor bisects the angle subtended by the three points and places a point along the bisector at the length specified. A negative length places the point along an obtuse angle bisector.

Construct Angle Intersection

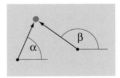

Given two points and two angles, this constructor places a point at the intersection of the rays defined by the points and angles.

Construct Deflection

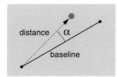

Given a line that serves as a baseline, a deflection angle, and a distance, this constructor places a point at the distance along a ray at the deflection angle.

Construct Deflection Intersection

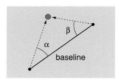

Given a line that serves as a baseline, a deflection angle measured from the start point of the baseline, and a deflection angle measured from the endpoint of the baseline, this constructor places a point at the intersection of the defined rays.

Construct Offset

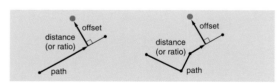

Given a path, a distance or ratio along the curve, and an offset distance, this constructor places a point at the offset. A positive value constructs a right offset and a negative value constructs a left offset. In civil engineering, this constructor is called a station offset.

Construct Parallel

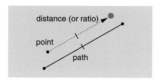

Given a path with straight lines, a reference point, and a distance, a new point is placed along a parallel curve. Note that the offset distance does not need to be specified; it is inferred from the geometry of the reference point and curve.

A similar polyline constructor, Contruct Offset, can create a parallel figure with paths containing circular arcs as well.

Multipoint construction

The constructors that follow return a set of points as a multipoint geometry.

Features with simple point geometries are far more common than features with multipoint geometries. When you apply these constructors, your application or tool may likely select one point from a multipoint, and then create or modify a simple point feature.

These constructors are present in the IConstructMultipoint interface implemented in the Multipoint class.

Construct Circular Arc Points

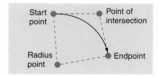

Given a circular arc, this constructor returns the endpoint, start point, radius point, and point of intersection. (Civil engineers refer to the start point as the point of curvature and the endpoint as the point of tangency.)

Construct Divide Equal

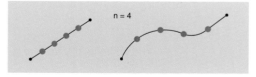

Given a curve and an integer number, this constructor places that number of points evenly spaced along the curve.

Construct Divide Length

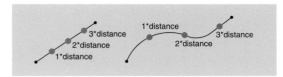

Given a curve and a length, this constructor places as many points as possible spaced by that length.

Construct Implied Intersection

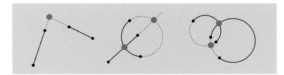

Given two segments, this constructor places points at the actual and extended intersections of those segments. A line segment is extended to an infinite ray, a circular arc is extended either tangentially or using its embedding circle, an elliptical arc is extended either tangentially or using its embedding ellipse, and a Bézier curve is extended tangentially.

Construct Intersection

Given two segments, this constructor places points only at the actual intersections of those segments.

Construct Tangent

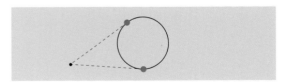

Given a circular arc and a point, this constructor places points at the positions of tangencies from the point to the arc.

Construct Three Point Resection

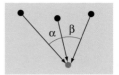

Given three points and two angles measured from a station point at unknown location, calculate and place the station point.

Line construction

A line is a straight segment between two points. Lines are building blocks for polylines and polygons.

This constructor is present in the IConstructLine interface implemented in the Line class.

Line ConstructAngleBisector

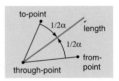

Given a from-point, through-point, and to-point and length, bisect that angle and construct a line at that length.

Circular arc construction

Circular arcs are segments that are part of the boundary of a circle. Like lines, circular arcs are building blocks for polylines and polygons.

Construct Arc Distance

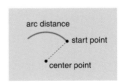

Given a center point, start point, and arc distance, this constructor builds a circular arc in a counterclockwise direction. The arc distance must be greater than zero. A Boolean value specifies whether the arc is constructed clockwise or counterclockwise.

Construct Chord Distance

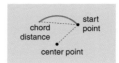

Given a center point, start point, and chord distance, this constructor builds a circular arc in a clockwise or counterclockwise direction, as specified by a Boolean value.

A chord is an imaginary line segment between the start point and endpoint of a circular arc.

Construct Chord Height

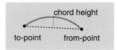

Given a from-point, to-point, and a height above the midpoint of a chord, this constructor builds a circular arc in the clockwise or counterclockwise direction, as specified by a Boolean value.

Construct Fillet

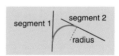

Given two segments and a radius length, this constructor builds a circular arc tangent to the two segments.

A fillet is a circular arc that is tangent to two segments. Most often, the tangent segments are lines, but they can also be circular arcs.

Construct Tangent and Point

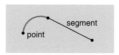

Given a segment and a point and a Boolean value specifying the start or endpoint of the segment, this constructor builds a circular arc.

Construct Three Points

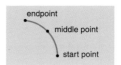

Given a start point, middle point, and endpoint, this constructor builds a circular arc that uniquely fits those points.

Construct Two Points and Radius

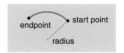

Given a start point, endpoint, and radius length, this constructor builds a circular arc. A Boolean value specifies whether the center point on the arc is on the left or right of the chord from the start point to the endpoint.

Curve construction

A curve is either a simple segment, a path, the boundary of a ring, a polyline, or the boundary of a polygon.

This constructor is present in the IConstructCurve interface that is implemented in the Polyline and Polygon classes.

Construct Offset

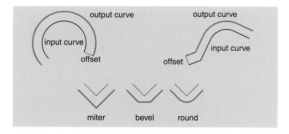

Given an input curve, this constructor constructs an offset figure that is offset by the specified distance.

You can specify whether the offset curve is mitered, beveled, or rounded. You can specify a beveling or rounding distance.

Path construction

A path is a sequence of connected segments.

This constructor is present in the IConstructPath interface that is implemented in the Path class.

Construct Rigid Stretch

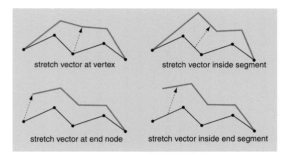

stretch vector at vertex   stretch vector inside segment

stretch vector at end node   stretch vector inside end segment

Given a path and a stretch vector, this creates a path that is proportionally stretched as shown.

This constructor is very useful for interactive rubber sheeting and conflation, which are editing techniques that adjust existing geometries to better conform to positions with higher accuracy, such as survey data.

Angle construction

These constructors return an angle and are present in the IConstructAngle interface implemented in the GeometryEnvironment class.

Construct Line

Given a line, calculate its angle.

Angle Construct Three Point

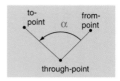

Given a from-point, through-point, and to-point, calculate the sweep angle between the points.

The ArcInfo geometry system includes a set of Boolean operators that test the spatial relationships between a base geometry and a comparison geometry. These operators can be applied to points, multipoints, polylines, and polygons.

The base geometry is the object invoking the operator. The comparison geometry is the geometry expressed as a parameter in the operator. The result of the relational operator is returned as a Boolean value. No new geometries are created with these operators.

Equals

Does the base equal the comparison geometry?

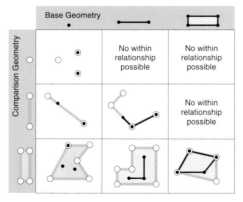

For the base geometry and comparison geometry to be equal, all of their constituent points must have identical coordinate values.

The geometries that are compared must have the same dimension.

Contains

Does the base geometry contain the comparison geometry?

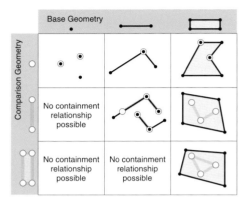

For the base geometry to contain the comparison geometry, it must be a superset of that geometry.

A geometry cannot contain another geometry of higher dimension.

Within

Is the base within the comparison geometry?

For the base geometry to be within the comparison geometry, it must be a subset of that geometry.

A geometry cannot be within another geometry of lower dimension.

Crosses

Does the base geometry cross the comparison geometry?

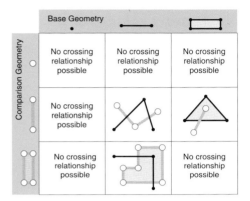

For the base geometry to cross a comparison geometry, they must intersect in a geometry of lesser dimension than the highest dimension.

Two lines can intersect at points. A line and an area can intersect at lines.

There is no crossing relationship possible between a base area and comparison area. This is considered an overlap relationship.

Disjoint

Is the base geometry disjoint from the comparison geometry?

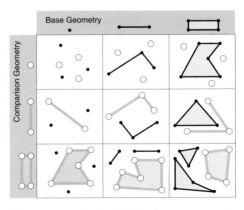

A base geometry is disjoint from a comparison geometry if they share no points.

Overlaps

Does the base geometry overlap the comparison geometry?

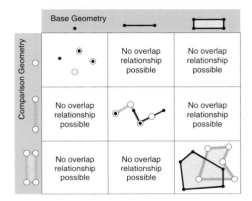

A base geometry overlaps a comparison geometry if their intersection is a geometry of the same dimension. An overlap relationship requires that both geometries be of the same dimension.

Touches

Does the base geometry touch the comparison geometry?

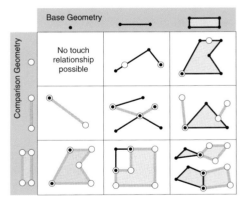

Two geometries touch when only their boundaries intersect.

The geometry system provides a set of operators that return geometries based on logical comparisons between sets of points in one or more geometries.

These operators provide support for editing geographic features that overlap. They are present in the ITopologicalOperator interface, which is implemented in the Envelope, Multipoint, Point, Polygon, and Polyline classes. In the GIS literature, these are sometimes called spatial topological operators.

Buffer

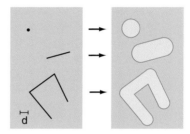

Given a geometry and a buffer distance, the buffer operator returns a polygon that covers all points whose distance from the geometry is less than or equal to the buffer distance.

Clip

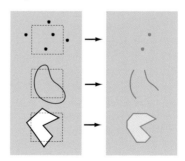

Given an input geometry and an envelope, the Clip operator returns a new geometry with the set of points of the input geometry that are within or on the boundary of the envelope.

Convex hull

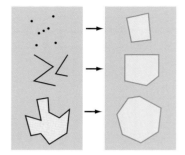

Given an input geometry, the convex hull operator returns a geometry that represents all points that are within all lines between all points in the input geometry.

A convex hull is the smallest polygon that wraps another geometry without any concave areas.

Cut

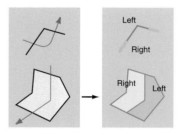

Given a cut curve and a geometry, the cut operator will split the geometry into a right part and left part, relative to the direction of the cut curve.

Points or multipoints cannot be split. Polylines and polygons must intersect the cut curve to be split.

Only two geometries are created by the cut operator, but they can have multiple parts.

Difference

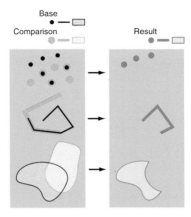

The difference operator returns a geometry that contains points that are in the base geometry and subtracts points that are in the comparison geometry.

Intersect

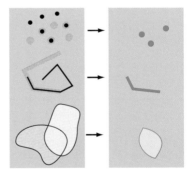

The intersect operator compares a base geometry (the object from which the operator is called) with another geometry of the same dimension and returns a geometry that contains the points that are in both the base geometry and the comparison geometry.

Symmetric difference

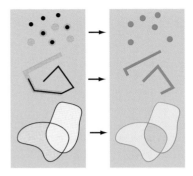

The symmetric difference operator compares a base geometry (the object from which the operator is called) with another geometry of the same dimension and returns a geometry that contains the points that are in the base geometry or the points in the comparison geometry, but excludes the points in both geometries.

Union

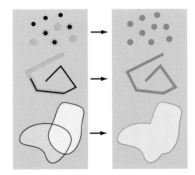

The union difference operator compares a base geometry (the object from which the operator is called) with another geometry of the same dimension and returns a geometry that contains the points that are in the base geometry together with the points in the comparison geometry.

Geometry object model

This is a simplified UML diagram extracted from the ArcInfo Object Model diagram that highlights the key geometry objects. This diagram is most useful for programmers, but also gives data modelers insight into the structure of feature shapes.

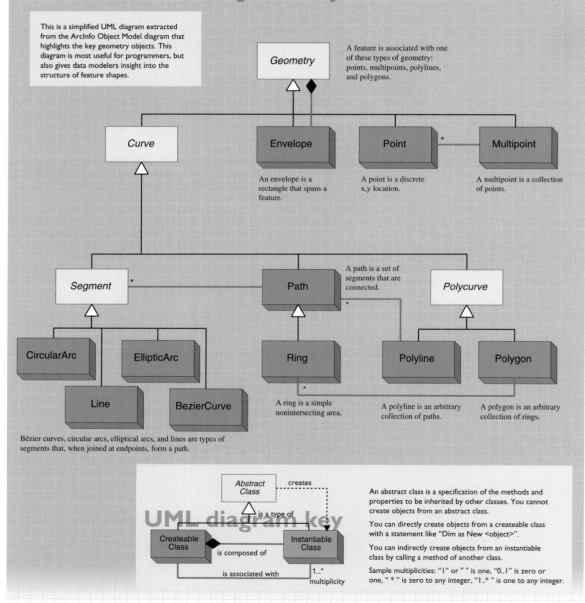

Geometry

A feature is associated with one of these types of geometry: points, multipoints, polylines, and polygons.

Curve

Envelope

An envelope is a rectangle that spans a feature.

Point

A point is a discrete x,y location.

Multipoint

A multipoint is a collection of points.

Segment

Path

A path is a set of segments that are connected.

Polycurve

CircularArc

EllipticArc

Line

BezierCurve

Ring

A ring is a simple nonintersecting area.

Polyline

A polyline is an arbitrary collection of paths.

Polygon

A polygon is an arbitrary collection of rings.

Bézier curves, circular arcs, elliptical arcs, and lines are types of segments that, when joined at endpoints, form a path.

UML diagram key

Abstract Class — creates — is a type of

Createable Class — is composed of

Instantiable Class

is associated with — 1..* multiplicity

An abstract class is a specification of the methods and properties to be inherited by other classes. You cannot create objects from an abstract class.

You can directly create objects from a createable class with a statement like "Dim as New <object>".

You can indirectly create objects from an instantiable class by calling a method of another class.

Sample multiplicities: "1" or " " is one, "0..1" is zero or one, " * " is zero to any integer, "1..* " is one to any integer.

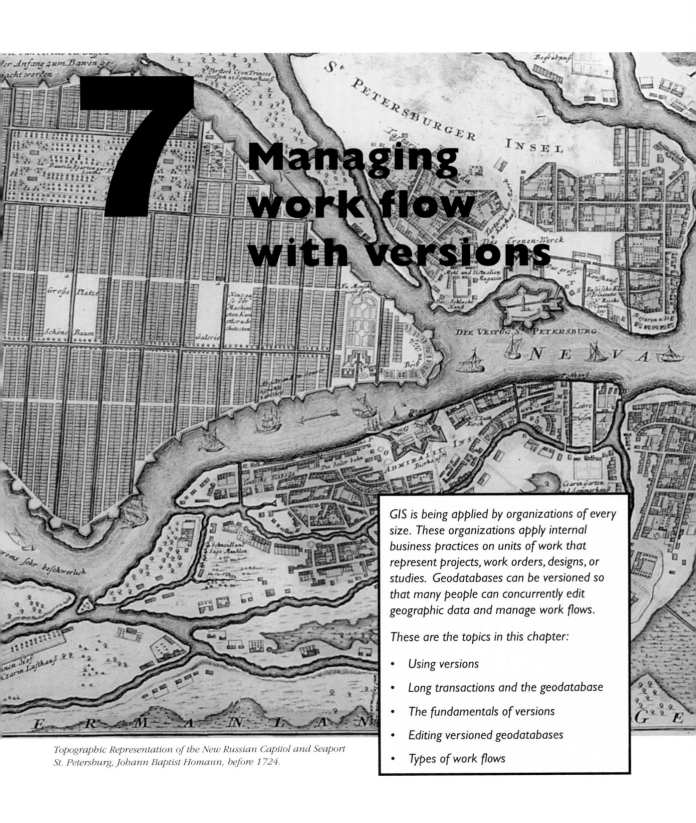

7

Managing work flow with versions

GIS is being applied by organizations of every size. These organizations apply internal business practices on units of work that represent projects, work orders, designs, or studies. Geodatabases can be versioned so that many people can concurrently edit geographic data and manage work flows.

These are the topics in this chapter:

- Using versions
- Long transactions and the geodatabase
- The fundamentals of versions
- Editing versioned geodatabases
- Types of work flows

Topographic Representation of the New Russian Capitol and Seaport St. Petersburg, Johann Baptist Homann, before 1724.

Many applications of GIS involve long-term design efforts that require the cooperation of a number of persons and departments. These design activities take place at the organizations that build things—utilities, municipal and regional governments, and departments of transportation.

These organizations have established work-flow processes for design, construction, and maintenance. The general steps include the initial engineering design, exploration of design alternatives, selection and approval of a design, the construction of the design, and updating maps with the construction features as they have been built in the field.

When you use GIS in these work-flow scenarios, it is necessary that multiple persons be able to simultaneously edit a geodatabase. They also need to have a transacted view of the geodatabase so that only the changes that they or their coworkers make are visible to them. Further, the work-flow structure needs to emulate the business practices of various departments in an organization.

The geodatabase data model serves these needs through a data management framework called *versioning*. This framework lets you create versions of a geodatabase for the states of a project, reconcile differences between versions, and update the master version of a geodatabase with the design as built.

This chapter documents the fundamentals of versioned geodatabases and shows how they can be employed with some work-flow scenarios.

DESIGN SCENARIO

To illustrate how versioned geodatabases are used in a multiuser environment, follow the scenario from a water utility shown on the facing page.

A municipal water utility keeps a comprehensive geodatabase with the current state of its field assets. All of the water pipes, valves, pumps, and other components of the water system are recorded as features in a geodatabase that is updated daily.

This water utility has a number of departments that are responsible for different phases of constructing and maintaining the water system. Because of this organizational structure, this utility uses a versioned geodatabase served through ArcSDE.

A versioned geodatabase has a top-level version that is always called "default." The default version of the geodatabase represents the water system in its best known as-built state. It is the starting point for creating new designs and construction activities.

Continuous editing of the geodatabase

The mapping department, Liz and Maria, are responsible for the daily maintenance of the geodatabase. To support the new line extension, Maria reviews field notes from that area and updates the water meter features. Liz adds new survey data points that were collected in a field survey in advance of the line extension. These edits are made directly to the default version of the geodatabase because they represent improved knowledge of the water system and are not part of a design cycle.

Creating versions by department

The information systems department is responsible for the corporate database that supports customer billing and asset management. To support the line extension, Fritz creates a new version, uses locators to match billing records to network features, and extracts and summarizes water usage data. This summarized data is intended only for this line extension project, so this version is temporary and discarded at the end of the design process.

The engineering department takes the data collected by the other departments and creates two versions for two engineering designs. Petra and Taylor create an engineering design based on using 16-inch pipe for the new main line. They simultaneously work on this version and create a proposed design. Filly creates another engineering design based on a 24-inch pipe to examine whether the increased pipe cost is offset by greater efficiency in handling present and future water usage. She discovers that the 24-inch pipe will serve projected water demand for 12 more years and that the greater initial construction cost is justified. Her design gets posted to the line extension version. When construction is complete, the line extension project version is posted into the default version.

This scenario is a simple example of how versions can be used to support a rich modeling environment for organizations that build complex systems.

Versioning scenario

The municipal water utility is planning an upgrade of a part of its system. Several departments are participating in data preparation and the design of two alternate line extensions. This scenario illustrates how a multilevel version tree lets many users collaborate on solving an engineering problem.

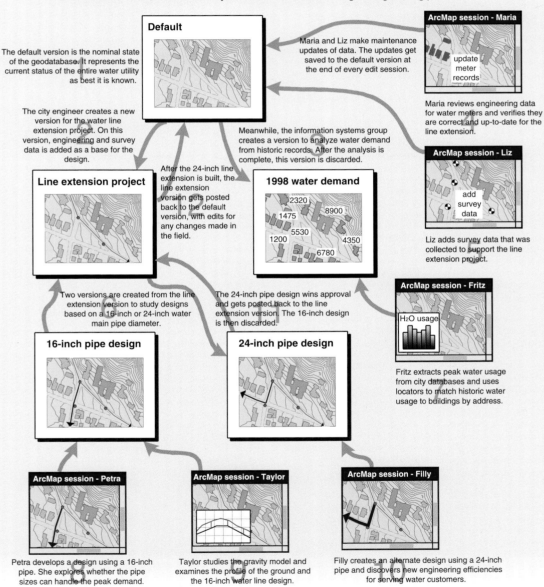

Default

The default version is the nominal state of the geodatabase. It represents the current status of the entire water utility as best it is known.

Maria and Liz make maintenance updates of data. The updates get saved to the default version at the end of every edit session.

ArcMap session - Maria

update meter records

Maria reviews engineering data for water meters and verifies they are correct and up-to-date for the line extension.

The city engineer creates a new version for the water line extension project. On this version, engineering and survey data is added as a base for the design.

Meanwhile, the information systems group creates a version to analyze water demand from historic records. After the analysis is complete, this version is discarded.

ArcMap session - Liz

add survey data

Liz adds survey data that was collected to support the line extension project.

Line extension project

After the 24-inch line extension is built, the line extension version gets posted back to the default version, with edits for any changes made in the field.

1998 water demand

2320
8900
1475
5530
1200
4350
6780

ArcMap session - Fritz

H₂O usage

Fritz extracts peak water usage from city databases and uses locators to match historic water usage to buildings by address.

Two versions are created from the line extension version to study designs based on a 16-inch or 24-inch water main pipe diameter.

The 24-inch pipe design wins approval and gets posted back to the line extension version. The 16-inch design is then discarded.

16-inch pipe design

24-inch pipe design

ArcMap session - Petra

Petra develops a design using a 16-inch pipe. She explores whether the pipe sizes can handle the peak demand.

ArcMap session - Taylor

Taylor studies the gravity model and examines the profile of the ground and the 16-inch water line design.

ArcMap session - Filly

Filly creates an alternate design using a 24-inch pipe and discovers new engineering efficiencies for serving water customers.

ArcInfo 8 is a milestone in the integration of GIS and relational database technology. GIS has now joined the mainstream of information technologies.

These are some of the benefits of storing all of your geographic data in commercial relational databases:

- You can integrate geographic data with corporate or agency databases.

- You can use standard database administration tools for managing your geographic data.

- You can create very large geographic databases that can be displayed and edited quickly.

- You can deploy geodatabases on the commercial relational database of your choice.

- You can serve geographic data to a wide variety of clients, such as view-only applications, CAD applications, or Internet applications.

The geodatabase extends standard relational (and object-relational) databases to support the special requirements of representing geographic data. These are some of the capabilities that a geodatabase adds to a relational database:

- You can represent and store geographic data in the form of raster datasets, feature datasets, TIN datasets, and location data.

- You can execute spatial and topological analysis on geographic data.

- You can perform rich cartographic display and produce high-quality maps.

- You can add intelligence to features by defining attributes, topological associations, relationships, and validation rules.

- You can enable many users to simultaneously display and edit geographic data.

This last capability, providing multiple users with read and write access to a geodatabase, is called *versioning* and is a critical requirement for many organizations. Versioning is a key function of geodatabases served through ArcSDE.

This chapter discusses how versioned geodatabases work with examples of work-flow scenarios.

TRANSACTIONS IN A DATABASE

A central idea of relational databases is a *transaction*. Simply put, a transaction is a group of atomic data operations that comprise a complete operational task.

Transactions preserve the consistency and integrity of the database by ensuring that either all or none of the atomic operations are executed for a task.

Short transactions

When you access data in a database, you have two basic goals: that the data be accurate and that it is timely.

Relational databases satisfy that requirement with *short transactions,* which represent operational tasks that can be completed in fractions of a second, or a minute or two at most. During the very brief time that a short transaction is being committed, no other updates to the affected rows are possible.

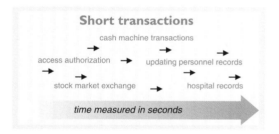

Short transactions represent most of the information tasks that people engage in, such as drawing money from an account at an automated teller machine, updating hours worked in a payroll application, or entering medical records.

Once a short transaction is committed in a relational database, it is not easy to undo that transaction or to reconstruct the state of the database at a historic point in time. There is only one state with a relational database: its status as of the most recently completed short transaction.

The short transaction model works very well for many critical applications that require instant access to a uniform view of data, but geographic data requires a longer view of updating data.

Editing geographic data

While a short transaction is in progress, the relational database applies locks on affected rows in database tables so that data being updated is protected from changes until the transaction is complete. When the short transaction is completed, the locks are released.

When multiple people are simultaneously editing geographic data, this type of row locking is impractical because even for short edit tasks, the locks must be held for several minutes.

Another reason that row locking is deficient for GIS is that features in a geodatabase coexist in a rich context of network connections, topological associations, and relationships.

To understand why, consider this scenario: You are editing an electric utility and adding lines, poles, transformers, and other devices. If another person were to edit a nearby feature while you were editing this transformer, your network could quickly fall into an inconsistent state.

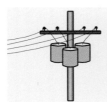

In a network, the integrity of one feature is dependent on others. When you edit a transformer, you must be certain that no one else is changing the pole it is mounted on, the voltage level of the circuit, and the line phasing of connected lines.

Short transactions fail for this scenario because when you are adding this transformer, you are not really editing a single feature, but you are editing a larger, more complex object—the network as a whole.

Another reason that short transactions are deficient in the multiuser geographic editing environment is that you must be able to always see the current state of the database as displayed on a map. Every time someone else made a change, your system would have to redraw the map, which may take a number of seconds for a complex map. This is unacceptable.

Long transactions

What you need for multiple users to edit geographic data is a transaction type that can do the following:

- Allows multiple persons to simultaneously edit the same complex system such as a network.

- Spans all of the edits that you need to perform on a work unit, whether it takes an hour or a month.

- Lets you have a private view of your data so that no one else sees incomplete work.

- Permits you to define the scope of work to match your business' work order system.

This type of transaction is a *long transaction*.

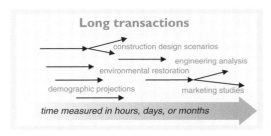

Long transactions

construction design scenarios
engineering analysis
environmental restoration
demographic projections
marketing studies

time measured in hours, days, or months

Long transactions have other uses besides representing construction work units. You can use long transactions to model any type of "what if" scenario.

During the scope of the long transaction, you can freely add proposed features, perform geographic analysis, and produce maps—all without affecting your nominal database. When the scenario is done, you can post the changes to the database if it is built or discard it if it is not.

Concurrency model

Long transactions implement a data management approach called *optimistic concurrency*. This means that when you start a long transaction, no locks are applied to features. The absence of locks permits the introduction of editing conflicts, but this is mitigated by an environment to easily detect, reconcile, and post these conflicts.

Optimistic concurrency is suitable for GIS applications because the volume of edits is small compared to the size of the geographic database. In real work-flow practices, edit conflicts are not frequent, and the cost of reconciling conflicts is minor when compared to the savings from not having to lock or check out features for the duration of a long transaction.

Versioning is ArcInfo software's implementation of long transactions against central multiuser relational databases served by ArcSDE. It is an advanced data management system that lets you adopt any of a variety of work-flow practices when editing geodatabases in a multiuser environment.

Versioning can be implemented on multiuser geodatabases served through ArcSDE. You cannot implement versioning on personal geodatabases.

BASIC CONCEPTS

Versioning lets multiple users directly edit a geodatabase without explicitly applying feature locks or duplicating data. The following are the essential facts about versions.

A version is a named state of a geodatabase

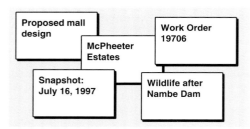

You can use versions to represent engineering designs, construction jobs, snapshots in time of geodatabases, and any type of scenario that involves the posing of "what if" questions in studying a result.

A version spans a geodatabase and has properties

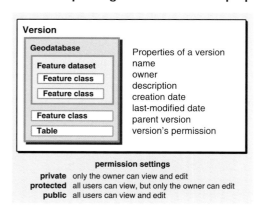

In ArcCatalog, you can define which objects in a geodatabase are versioned. You can selectively specify which feature datasets, feature classes, and tables are versioned.

When you specify that a feature dataset is versioned, all of its tables and feature classes are automatically versioned.

You can control the visibility of a version to other users by setting its permission.

A geodatabase can have multiple coexisting versions

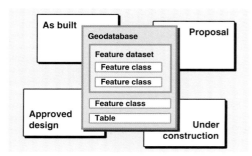

Each version lets you perform all of the same display and analytic functions as a nonversioned geodatabase.

Versions differ from each other only in row state, not in schema

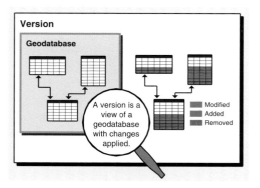

A version presents you with a seamless view of all the edits applied since the version was created. The row state reflects all added, removed, and modified objects. The row state information about each version is stored (or persisted) in the geodatabase. The schema, the definition of tables and their fields, can be modified on a geodatabase; schema changes are applied to all versions of the geodatabase.

Internally, the geodatabase has tables to keep track of modified, added, and removed features for each version, but that is not apparent to you when you use versions. It appears to be an integral copy of a geodatabase.

Every versioned geodatabase has a default version

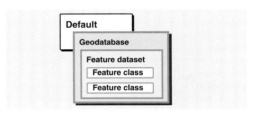

The default version can be thought of as the "as-built" version. It usually represents the nominal state of the geodatabase.

The default version is the geodatabase. Most users will edit the default version.

A version is created from another version

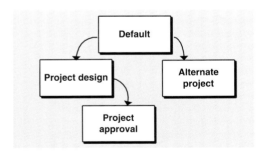

Starting with the default version, you can create any number of versions. Every version, except for the default version, has exactly one parent version. You can create a complex hierarchical version tree as appropriate for your organization's work-flow requirements.

Versions can be removed, but only if their child versions are removed first. Before removing a version, the changes can be reconciled and posted to another version or discarded.

A versioned geodatabase can be compressed. Over time, rows are added to various internal tables that manage versions in a geodatabase and many are superseded by other rows. These extra rows can be eliminated to conserve disk space and preserve data access performance. This is a task for the ArcSDE administrator.

A user can connect to any version

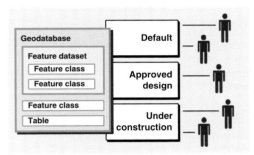

A user will start editing a version based on the project or project stage they are working on. A user can work on any version they have been granted permission to.

WHY VERSIONS PERFORM WELL

When you start using versioning, you will notice considerable performance improvement and greater ease of use over previous data management systems, such as checked out datasets, tiled libraries of datasets, or copied datasets.

The reason that versioning works quickly and well is that versions do not require any duplication or replication of data. Internally, a versioned geodatabase uses internal identifiers and manages additional tables that record which features and objects are added, removed, or modified.

It is not necessary for a data modeler, or even a programmer or database administrator, to be aware of any of the implementation details of versioned geodatabases. ArcCatalog, ArcMap, and ArcSDE collectively provide you with an easy-to-use and comprehensive user interface to versioning.

The ArcMap Editor is the arena in which you perform operations on versions in accordance with your organization's work-flow practices.

The basic versioning operations you can do in ArcMap are editing a version, reconciling your edit session against another version and resolving any conflicts that arise, and then posting the changes in your edit session to a version.

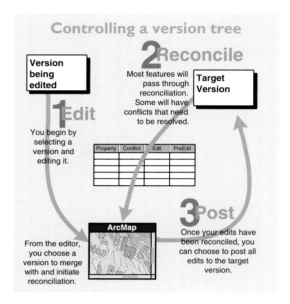

Editing a version

When you are using ArcMap and start editing, if your map references one version, then that version is automatically opened for editing. If your map references multiple versions, you can choose from one of these to start editing.

Reconciling versions

Reconciliation is the process of merging features and objects from a target version into the current edit session. A target version can be any version in the direct ancestry of the version being edited. Reconciliation must be done before posting changes to another version.

When you reconcile, you should have full permission to modify the feature classes in the target version that you have been editing in the edit

session. If you do not, you will receive an error message and will not be able to complete the reconciliation.

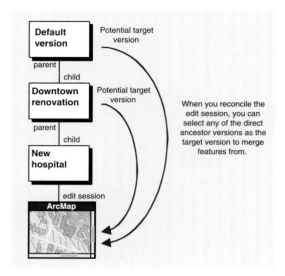

Because a version spans all the versioned feature datasets, feature classes, and tables in a geodatabase, all objects and features in these classes will be merged into the edit session. The great majority of features and objects will pass straight through reconciliation from the target version to your edit session.

Handling conflicts during reconciliation

A small percentage of features and objects will have conflicts when compared between the target version and the edit session.

There are two types of conflicts:

1 When the same feature is updated in both the target version and the edit session.

2 When the same feature is updated in one version and deleted in the other.

For most reconcile operations, no conflicts will be encountered. That is because at most organizations, projects and versions represent distinct geographic areas. If you and your coworkers are editing different parts of the map, it is generally not possible to introduce conflicts. Conflicts usually arise when multiple people are editing features that are in close proximity.

When conflicts do arise, you will see an interactive conflict resolution dialog. This dialog lets you examine and zoom to any conflicts between the two versions. The conflicts are grouped by *conflict classes,* which are feature classes and tables that conflicts are detected for.

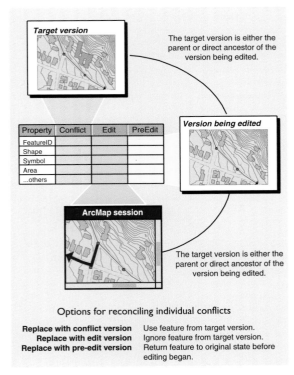

The target version is either the parent or direct ancestor of the version being edited.

The target version is either the parent or direct ancestor of the version being edited.

Options for reconciling individual conflicts

Replace with conflict version   Use feature from target version.
Replace with edit version   Ignore feature from target version.
Replace with pre-edit version   Return feature to original state before editing began.

For each conflict, you can choose whether to replace the feature in your edit session with the conflict feature from the target version, keep it as it is in your edit session, or revert it to its state at the beginning of your edit session.

When you work with features that have relationships with features in other feature classes, there are additional considerations for deciding how to resolve conflicts. For details, read chapter 11, "Working with a versioned geodatabase," in *Building a Geodatabase.*

Posting versions

You can post a version to the target version after you have successfully completed the reconciliation.

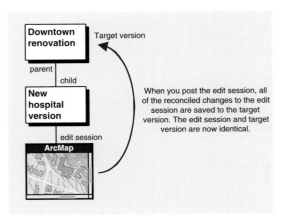

When you post the edit session, all of the reconciled changes to the edit session are saved to the target version. The edit session and target version are now identical.

The post operation synchronizes the row state of your edit session with the target version. They are identical at this point.

At this point, you may continue to make more edits in your edit session, but you will need to undergo the reconciliation, conflict resolution, and posting process again if you want to apply these changes to the target version.

If a posting marks the end of your project, you can go to ArcCatalog to terminate that part of your work flow by removing the version you have been editing.

When you apply versioned geodatabases to your organization, you can select from one or several types of work flows that match your business practices.

The following are a summary of the basic work flows supported by versioning. Your implementation can be one or a combination of these work flows.

Direct editing

The simplest work flow for multiuser access on a geodatabase is for many users to directly edit the default version.

As each person opens the default version for editing, a temporary version is created. The editor is not explicitly aware that a version is created and does not give it a name. Whenever the editor saves the work or ends the edit session, then that temporary version is automatically reconciled with and posted to the default version.

If there are conflicts, you must resolve them with the conflict resolution dialog before you can successfully save your edits. If no conflicts are detected, the edits are directly posted to the default version.

This work flow has the virtue of simplicity. It is most appropriate where the units of work are fairly modest in scale and where no design alternatives have been explored or historical snapshots made.

Two-level tree

Many organizations employ a more structured process that tracks discrete work units of construction or maintenance.

These work units typically span a time interval of days, weeks, or months and represent tasks such as adding new phone service, adding a new line extension with pipes or poles and wires, and building a new pump station or electric substation.

When a work order or project is initiated, a version is created. One or several people work on this version until the design or construction is complete. At that point, reconciliation and posting are done to merge the work order features into the default versions, and then the work order version can be removed.

Multilevel tree

Some organizations' projects have a higher level of structure and can be subdivided into functional or geographic parts.

For example, a project to design and construct a new shopping mall might have phases of construction, be subdivided into east and west parts, or be subdivided by construction activities such as structure, gas and water, and electric.

For larger projects with departments and teams, a multilevel version tree is an effective way to organize work flow. The teams that are working on each aspect of the project have their own version, with which they can maintain a private view of their designs and then post the designs when constructed.

Cyclical

Many projects go through a prescribed or regulated set of stages that require engineering, administrative, or legal approval before proceeding to the next stage.

A version represents each stage of this process. A cyclical work flow can capture the design at each stage, and when the last stage is reached and finished, the design can be posted directly to the default version, which represents the nominal state of the database.

This work flow saves the effort of progressively posting changes up the version tree; you can bypass the immediate parent versions and post directly to the default or other version.

Extended history

For some projects, it is desirable to preserve a version that reflects a historic state of a project.

You can define a historic version on a project version, and when the project version is posted to its parent version, the historic version remains as a snapshot in time.

SUMMARY

In practice, you will probably either apply the direct editing work flow or some combination of the others. An understanding of the elements of work-flow management will improve the effectiveness of your geodatabase design.

Work flows with a versioned geodatabase

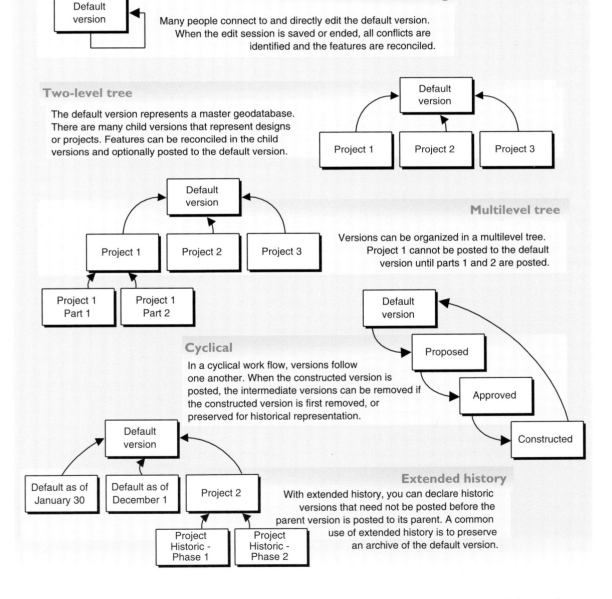

Direct editing

Default version

Many people connect to and directly edit the default version. When the edit session is saved or ended, all conflicts are identified and the features are reconciled.

Two-level tree

The default version represents a master geodatabase. There are many child versions that represent designs or projects. Features can be reconciled in the child versions and optionally posted to the default version.

Default version

Project 1   Project 2   Project 3

Default version

Project 1   Project 2   Project 3

Project 1 Part 1   Project 1 Part 2

Multilevel tree

Versions can be organized in a multilevel tree. Project 1 cannot be posted to the default version until parts 1 and 2 are posted.

Default version

Proposed

Approved

Constructed

Cyclical

In a cyclical work flow, versions follow one another. When the constructed version is posted, the intermediate versions can be removed if the constructed version is first removed, or preserved for historical representation.

Default version

Default as of January 30   Default as of December 1   Project 2

Project Historic - Phase 1   Project Historic - Phase 2

Extended history

With extended history, you can declare historic versions that need not be posted before the parent version is posted to its parent. A common use of extended history is to preserve an archive of the default version.

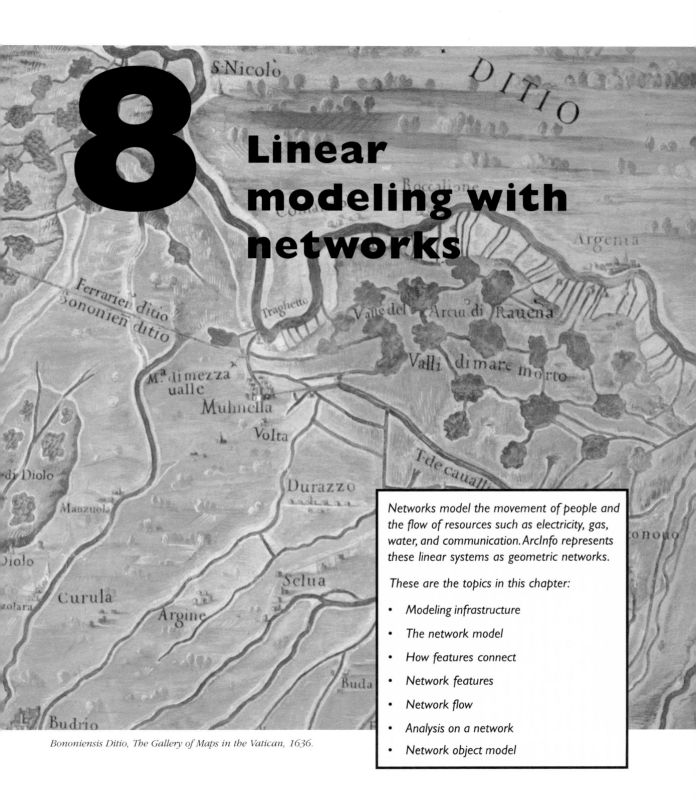

8

Linear modeling with networks

Bononiensis Ditio, The Gallery of Maps in the Vatican, 1636.

Networks model the movement of people and the flow of resources such as electricity, gas, water, and communication. ArcInfo represents these linear systems as geometric networks.

These are the topics in this chapter:

- Modeling infrastructure
- The network model
- How features connect
- Network features
- Network flow
- Analysis on a network
- Network object model

The economic foundation of our world is its infrastructure: the collection of highways, cables, and pipelines that enables the movement of people, energy, commodities, and ideas.

This infrastructure is modeled as networks, and the form, capacity, and efficiency of these networks have a substantial impact on our standard of living and our perception of the world around us.

ArcInfo 8 introduces a new network model—the geometric network—that builds on years of experience modeling transportation and utility networks. Geometric networks let you reach a new level of sophistication for naturally modeling infrastructure.

These are the principal benefits of the geometric network model:

- Editing networks is easy. When you add network features, you can ensure that they are properly connected with network connectivity rules.

- Network features can represent complex parts of a network, such as switches. This simplifies editing and lets you create better maps with fewer features in your network representation.

- A suite of simple and advanced network analysis solvers is built into ArcInfo, ready to use. Network analysis is fast even on very large datasets.

- Networks can be versioned. Many people can simultaneously edit the same large network in compliance with their organization's work-flow practices.

This chapter documents the important qualities of networks and reveals how the geometric network and the underlying logical network form the basis for advanced modeling of transportation, energy, and communication systems.

NETWORKS AND APPLICATIONS

Networks are simple. They are comprised of two fundamental components, *edges* and *junctions*.

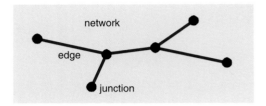

Some examples of edges are streets, transmission lines, pipe, and stream reaches.

Some examples of junctions are street intersections, fuses, switches, service taps, and the confluence of stream reaches.

Edges connect at junctions and the flow from one edge—automobiles, electrons, water—can be transferred to another edge.

From this simple idea, you can build networks to serve any of a myriad of applications. Here are a few examples:

- A railroad schedules its trains to efficiently link with intermodal container trucks.

- A parcel delivery service optimizes its package delivery on a street system.

- An electric utility locates where power outages originate based on telephone calls received from affected customers.

- An environmental agency analyzes water samples collected from streams to trace contaminant flow.

- A regional transportation agency uses traffic data to plan future highway construction.

- A school district finds optimum bus routes to pick up children and deliver them to school.

- A driver uses a mapping system with a GPS receiver mounted in the car to find the best way to get to a destination.

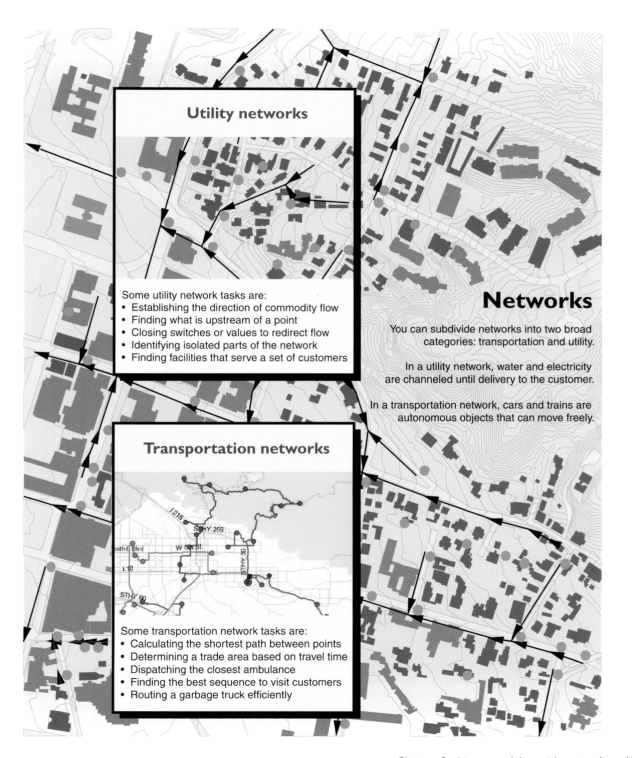

Utility networks

Some utility network tasks are:
• Establishing the direction of commodity flow
• Finding what is upstream of a point
• Closing switches or values to redirect flow
• Identifying isolated parts of the network
• Finding facilities that serve a set of customers

Networks

You can subdivide networks into two broad categories: transportation and utility.

In a utility network, water and electricity are channeled until delivery to the customer.

In a transportation network, cars and trains are autonomous objects that can move freely.

Transportation networks

Some transportation network tasks are:
• Calculating the shortest path between points
• Determining a trade area based on travel time
• Dispatching the closest ambulance
• Finding the best sequence to visit customers
• Routing a garbage truck efficiently

The geodatabase has a dual representation of a linear system—the geometric network and the logical network.

The *geometric network* is the set of features that participate in a linear system. The geometric network matches a view of a network as a collection of features.

A geometric network is associated with a *logical network,* which is a pure network graph consisting of edges and junction *elements.*

Together, these two representations of a network provide a rich model for storing and analyzing linear systems.

THE GEOMETRIC NETWORK

A geometric network is a collection of features that comprise a connected system of *edges* and *junctions.* An edge has two junctions and a junction can be connected to any number of edges.

Edge features can cross in two-dimensional space without intersecting. An example is a bridge over a road. This is called *nonplanarity.*

The features that represent edges and junctions are called *network features.* Only network features can participate in a geometric network.

A *network feature class* is a homogeneous collection of one of these types of network features: *simple junction feature, complex junction feature, simple edge feature,* or *complex edge feature.*

More than one network feature class can represent a given topological role in a geometric network. A network feature class is associated with exactly one geometric network.

Network features in a geometric network have all the same characteristics as other features:

- You can create as many feature classes as necessary for edges and junctions. You can add any attributes to these feature classes.

- You can define subtypes for major feature classifications and apply default values, attribute domains, and split/merge policies on attributes.

- You can establish relationships among network features and any other feature or object.

- For advanced applications, you can extend a network feature class and create custom network features.

Network features have additional specialized behaviors that preserve connectivity and automatically update network elements.

THE LOGICAL NETWORK

Like a geometric network, a *logical network* is a collection of connected edges and junctions. The key difference is that a logical network does not have coordinate values. Its main purpose is to store the connectivity information of a network along with certain attributes.

Since edges and junctions in a logical network contain no geometry, they are not features, but *elements.* There is a one-to-one or one-to-many relationship between network features in a geometric network and network elements in a logical network.

A geometric network is always associated with a logical network. The logical network elements are automatically updated when you edit network features.

The logical network does not directly appear in the ArcInfo applications. Rather, you interact with the geometric network. The logical network is the basis of the sophisticated behavior of the network features.

Reading the diagrams

There are conceptual diagrams throughout this chapter that show the relationship between the geometric network and its logical network.

Many of the details about the logical network are simplified—it is not necessary for the data modeler to know all of the internal implementation.

While the logical network is invisible in the ArcInfo applications, understanding its basic concepts will give you insight into building network models.

Two views of a network

You can view a network as a collection of geographic objects such as rails, roads, stations, and bridges and also as a pure network of edges and junctions.

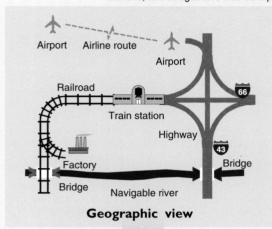

Geographic view

Network view

A geometric network is the representation of geographic features that comprise a network.

A logical network is a pure graph of junction elements and edge elements.

Geometric network

The geometric network maintains relationships between connected junction features and edge features. When you move a junction feature, the connected edge features are rubberbanded.

Network features can be organized in any number of network feature classes.

A network feature can be related to one or many network elements. This allows a single feature to represent a complex part of a network.

You can define connectivity rules to define the valid combinations of connected junctions and edges in a geometric network.

Logical network

Network elements are stored in an edge table and junction table with a connectivity table describing how they connect.

A logical network has no geometry or coordinates. It is a pure graph of how junctions and edges are connected.

The connectivity table keeps track of how edge and junction elements are connected.

The logical network is invisible in ArcMap and ArcCatalog, but it is the foundation for the geometric network's rich model and high performance.

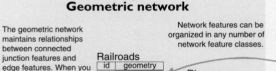

Features — **Elements**

Features from an arbitrary number of edge and junction feature classes correspond to network elements in the edge table and junction table.

You interact with the network through network features. When you add or remove a network feature in a geometric network, ArcInfo adds or removes the matching network elements. When you perform network analysis, ArcInfo submits a solver to the logical network.

The geometric network and logical network are always synchronized.

The centerpiece of a logical network is the connectivity table, which describes how network elements are connected.

The logical network also contains a junction element table and an edge element table.

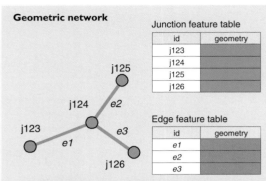

Geometric network

Junction feature table

id	geometry
j123	
j124	
j125	
j126	

Edge feature table

id	geometry
e1	
e2	
e3	

A geometric network contains the features that participate in a network. Feature classes contain either edge features or junction features.

Logical network

Connectivity table

Junction	Adjacent junction and edge		
j123	j124, e1		
j124	j124, e1	j125, e2	j126, e3
j125	j124, e2		
j126	j124, e3		

A logical network contains the connectivity of the network. The connectivity table lists all the adjacent junctions to a given junction, along with the edge that connects them.

For every junction in the network, the connectivity table lists the adjacent junctions and edges—junctions at the other end of the connected edge.

The connectivity table is how the geometric network maintains the integrity of the network.

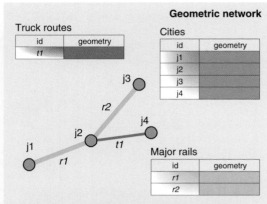

Geometric network

Truck routes

id	geometry
t1	

Cities

id	geometry
j1	
j2	
j3	
j4	

Major rails

id	geometry
r1	
r2	

A geometric network can have any number of participating feature classes. In this example, there is one junction feature class (Cities) and two edge feature classes that connect the junctions (Major rails and Truck routes).

Logical network

Junction element table

Feature Class	Feature ID	Element ID
1	j1	0
1	j2	1
1	j3	2
1	j4	3

Edge element table

Feature class	Feature ID	Element ID
2	r1	10
2	r2	11
3	t1	12

Connectivity table

Junction	Adjacent junction and edge elements		
0	1, 10		
1	0, 10	2, 11	3, 12
2	1, 11		
3	1, 12		

The logical network tracks feature IDs by feature class. For each feature class and feature ID combination, the logical network creates its own internal element ID.

The junction element and edge element tables provide a unique element ID that is a combination of the feature class and the feature ID.

CONNECTIVITY RULES

In most networks, not all edges can connect to all other junctions. Also, not all edges can connect to all other edges through a specified junction. For example, a hydrant lateral in a water network can connect to a hydrant, but not to a service lateral. Similarly, a 10-inch transmission main can only connect to an 8-inch transmission main through a reducer.

Network connectivity rules constrain the type of network features that may be connected to one another, and the number of features of any particular type that can be connected to features of another type.

Connectivity rules let you easily maintain the integrity of the network features in a geometric network. At any time, you can selectively validate features in the database and generate reports as to which features in the network are violating one of the connectivity or other rules. The following are the connectivity rules for network features.

Edge–junction rule

This rule constrains which types of junctions can connect to a type of edge.

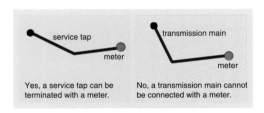

Yes, a service tap can be terminated with a meter.

No, a transmission main cannot be connected with a meter.

Meters can only connect to low-voltage lines.

Edge–edge rule

This rule establishes which combinations of edge types can connect through a given junction.

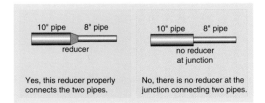

Yes, this reducer properly connects the two pipes.

No, there is no reducer at the junction connecting two pipes.

Two pipes of different diameters can be connected only through a properly sized reducer.

Edge–junction cardinality

This rule lets you restrict the count (cardinality) of edges that connect at a junction.

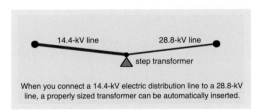

Yes, a switch can connect two through four lines.

No, a switch cannot be the terminus of a line.

A certain type of switch might be designed to accept between two and four lines. You can precisely define the acceptable range of lines that can be connected at a junction.

Default junction type

When you connect one type of edge to another, you can specify a default junction type to be inserted.

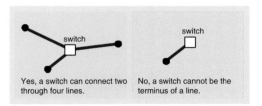

When you connect a 14.4-kV electric distribution line to a 28.8-kV line, a properly sized transformer can be automatically inserted.

When a 14.4-kV line is added to an end-junction of a 28.8-kV line, a step-down transformer with the correct electrical ratings is assigned to the junction.

Features can play four roles in a geometric network: simple edge, simple junction, complex edge, and complex junction.

Each feature class in a geometric network contains features of one of these types. A geometric network can have multiple feature classes with the same role.

This is a simplified portion of the geodatabase data access model.

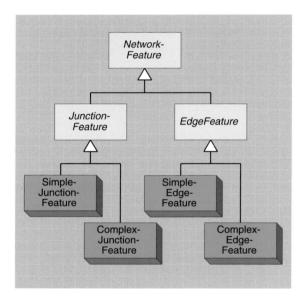

There are two types of network features: junction and edge. There are two types of junction features: simple and complex. There are two types of edge features: simple and complex.

SUMMARY OF NETWORK FEATURES

A *simple edge feature* is associated with a single edge in a logical network.

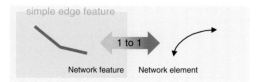

A *simple junction feature* is a feature associated with a single node in a logical network.

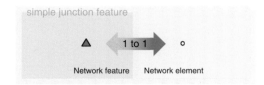

A *complex edge feature* is associated with any number of edges in a logical network. These edges must be arranged in a chain configuration.

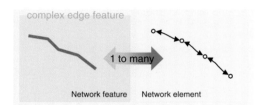

A *complex junction feature* is associated with a collection of junctions and edges in a logical network. The edges and junctions are connected and may be arranged in any topological configuration. These elements can be considered to be an internal network represented by a complex junction feature.

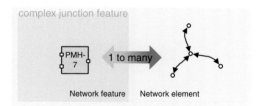

The complex junction feature must be implemented by writing a custom feature type. You cannot create a new feature class based on complex junction features without writing software code compliant with the ArcInfo class extension framework.

Simple edge and junction features

These network features have a one-to-one correspondence with network elements. They are suitable for simple parts of networks, but the edge-splitting scenario shows one limitation of simple network features.

Simple edge and junction features

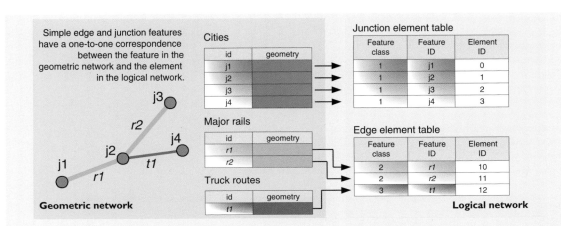

Simple edge and junction features have a one-to-one correspondence between the feature in the geometric network and the element in the logical network.

Geometric network

Cities

id	geometry
j1	
j2	
j3	
j4	

Major rails

id	geometry
r1	
r2	

Truck routes

id	geometry
t1	

Junction element table

Feature class	Feature ID	Element ID
1	j1	0
1	j2	1
1	j3	2
1	j4	3

Edge element table

Feature class	Feature ID	Element ID
2	r1	10
2	r2	11
3	t1	12

Logical network

Splitting a simple edge

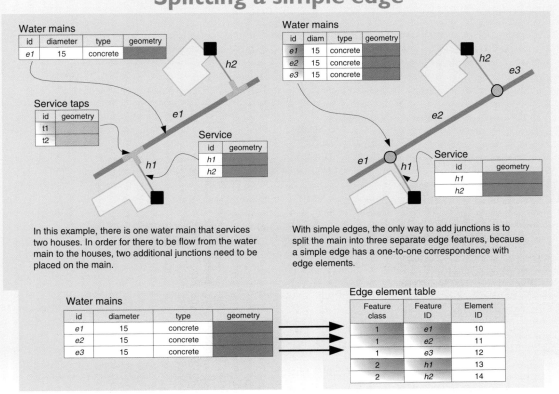

Water mains

id	diameter	type	geometry
e1	15	concrete	

Service taps

id	geometry
t1	
t2	

Service

id	geometry
h1	
h2	

In this example, there is one water main that services two houses. In order for there to be flow from the water main to the houses, two additional junctions need to be placed on the main.

Water mains

id	diam	type	geometry
e1	15	concrete	
e2	15	concrete	
e3	15	concrete	

Service

id	geometry
h1	
h2	

With simple edges, the only way to add junctions is to split the main into three separate edge features, because a simple edge has a one-to-one correspondence with edge elements.

Water mains

id	diameter	type	geometry
e1	15	concrete	
e2	15	concrete	
e3	15	concrete	

Edge element table

Feature class	Feature ID	Element ID
1	e1	10
1	e2	11
1	e3	12
2	h1	13
2	h2	14

Complex edge features

Imagine that your data has a municipal water main that runs for several hundred meters along a street. Along this main are service taps (junctions) connected to household service pipes. All your data query and maintenance functions ideally need to treat this main as one single feature.

But because your network analysis functions need to model flow from the main to the services, the logical network needs to treat the portions between each service tap as a single edge. Using simple edges, this main would have to be broken into sections between each junction. So instead of

having one long main, there are now many fragmented sections of the main, which complicates data query and maintenance.

With simple edges, a single feature has to be split into many features in order to connect other features with it. This may be undesirable for many databases, leading to fragmented databases and complex rules of behavior.

Complex edges solve the fragmentation problem by allowing junctions to be placed anywhere along their length without creating new edge features. A geometric network with complex edge features creates many edge elements for each feature.

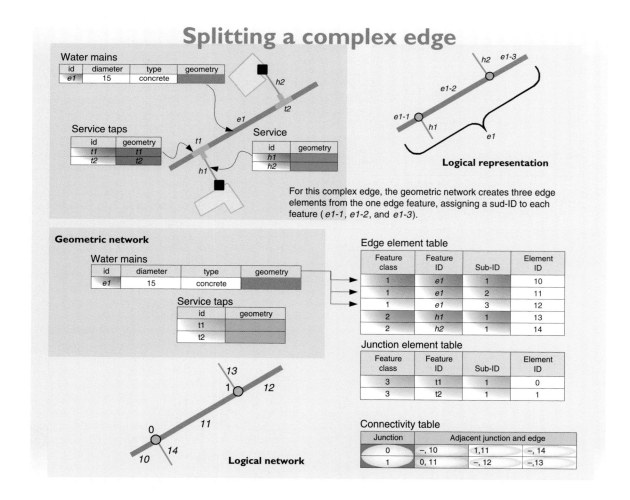

For this complex edge, the geometric network creates three edge elements from the one edge feature, assigning a sud-ID to each feature (e1-1, e1-2, and e1-3).

Complex junction feature

A complex junction feature corresponds to any number of edge and junction elements in the logical network.

The best way to understand complex junctions is to imagine a switch cabinet in an electrical network. Switch cabinets are actually miniature networks, consisting of their own simple junctions and edges.

A complex junction is ideal for modeling networks within networks, as in the case of electrical switches. A complex junction can contain any number of edges and junctions.

Complex junction features

Schematic

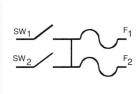

Complex junctions are used in electrical networks to represent complex switches. This schematic of a complex switch shows two simple switches (SW) and two fuses (F).

Logical network

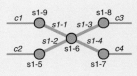

In the logical network, this switch is modeled with four edge elements and five junction elements. You can write code in a custom switch object to manage the edges and junction elements in the logical network.

Edge element table

Feature class	Feature ID	Sub-ID	Element ID
1	s1	1	10
1	s1	2	11
1	s1	3	12
1	s1	4	13
2	c1	1	14
2	c2	1	15
2	c3	1	16
2	c4	1	17

Junction element table

Feature class	Feature ID	Sub-ID	Element ID
-	s1	5	20
-	s1	6	21
-	s1	7	22
-	s1	8	23
-	s1	9	24

Geometric network

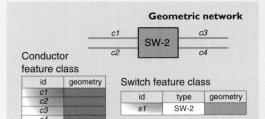

Conductor feature class

id	geometry
c1	
c2	
c3	
c4	

Switch feature class

id	type	geometry
s1	SW-2	

In a geometric network, this switch is displayed as a box labeled with the type of switch, in this case "SW-2." Two wires enter and leave this switch. This switch could be implemented as a complex junction feature.

Pump station scenario

This example shows a pump station as a complex junction. The pump station is displayed as a box. Inside this box is a miniature network containing three valves, a check valve, a pump, a meter, and a tee, all modeled as simple junctions within the one complex junction.

In ArcInfo, a developer creates complex junctions as a custom feature, implementing the rules on how the junction is stored in the logical network.

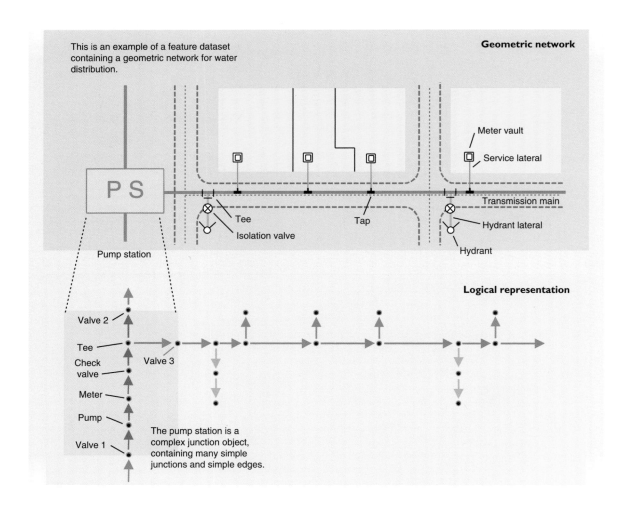

Networks fall neatly into one of two operational contexts, utility network or transportation network. In a transportation network, the commodities that flow through the network—automobiles—have a "will of their own." The driver of the automobile decides how they will flow through the network. In a utility network, the commodity that flows through the network—water, electricity, oil—has no will of its own. The network imposes flow direction by its configuration of sources, sinks, and switches.

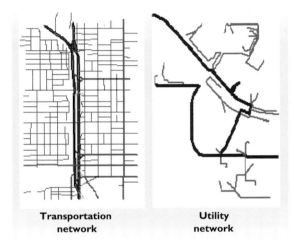

Transportation network

Utility network

In utility network applications, the direction of commodity flow along edges needs to be an intrinsic part of the network. For example, it is usually a bad thing to have water flowing in from both ends of a solitary pipe—the water is either going to back up or the pipe is going to burst.

If the geometric network is used for operational decision making, such as whether to close a switch or open a valve, you have to know if the decision will result in incorrect flow. In analysis, it is usually a requirement to know what features are downstream (with the flow) or upstream (against the flow) of some location.

A geometric network has a method to establish flow direction. This method decides how commodities flow in the network based on the current configuration of sources and sinks and the enabled state of each feature. The result of this method is to align the direction that commodities flow along

each edge, either with the direction of the feature or against the direction of the feature, relative to its digitized direction.

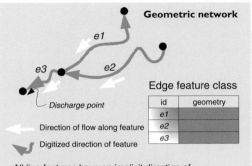

Geometric network

Discharge point

← Direction of flow along feature

↘ Digitized direction of feature

Edge feature class

id	geometry
e1	
e2	
e3	

All line features have an implicit direction of digitization, which is the x,y coordinate order. This is an example of a simple stream network. All water flows to the discharge point. Water flows opposite the digitized direction of feature *e1*, but with the digitized direction of features *e2* and *e3*.

Logical network

Edge element table

Feature class	Feature ID	Sub-ID	Element ID	Flow direction
1	e1	1	10	Against
1	e2	1	11	With
1	e3	1	12	With

The establish flow direction method on a geometric network populates the flow direction property of each edge element in the logical network. There are two possible values: flow is against digitized direction or with digitized direction. This flow direction information is critical for many applications and is used when flow direction is to be displayed on a geometric network.

SOURCES AND SINKS

In a utility network, sources and sinks are used in determining flow direction. Any junction feature class can take on the ancillary role of a source or a sink. A *source* is a junction from which a commodity flows, such as a well-head pump. A *sink* is a junction where all commodity flow terminates, such as a wastewater treatment plant.

When you build a geometric network, you say whether or not features in a junction feature class can assume this ancillary role. If they can, the editor can be used to specify whether an individual junction within the feature class is either a source or sink.

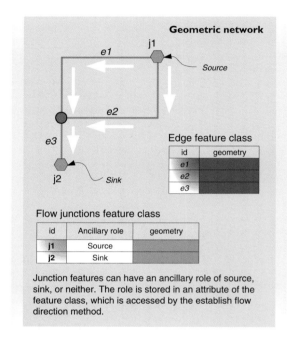

Geometric network

Source

Sink

Edge feature class

id	geometry
e1	
e2	
e3	

Flow junctions feature class

id	Ancillary role	geometry
j1	Source	
j2	Sink	

Junction features can have an ancillary role of source, sink, or neither. The role is stored in an attribute of the feature class, which is accessed by the establish flow direction method.

The ancillary role field determines if a junction feature and element are a source or sink.

DISABLED FEATURES

All features participating in a network have an *enabled/disabled state*. Features that are disabled do not participate in network flow: nothing flows into or out of the feature. Disabled features are useful for representing open electrical switches or closed valves.

Sources, sinks, and the enabled/disabled state all affect how flow is established in a network.

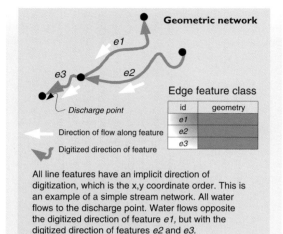

Geometric network

Discharge point

Direction of flow along feature

Digitized direction of feature

Edge feature class

id	geometry
e1	
e2	
e3	

All line features have an implicit direction of digitization, which is the x,y coordinate order. This is an example of a simple stream network. All water flows to the discharge point. Water flows opposite the digitized direction of feature *e1*, but with the digitized direction of features *e2* and *e3*.

Logical network

Edge element table

Feature class	Feature ID	Sub-ID	Element ID	Flow direction
1	e1	1	10	Against
1	e2	1	11	With
1	e3	1	12	With

The establish flow direction method on a geometric network populates the flow direction property of each edge element in the logical network. There are two possible values: flow is against digitized direction or with digitized direction. This flow direction information is critical for many applications and is used when flow direction is to be displayed on a geometric network.

INDETERMINATE FLOW

It may not be possible to establish flow direction for an edge. This only occurs when the sources, sinks, and disabled features do not give enough information. When flow direction cannot be established for an edge, it has *indeterminate flow*.

Indeterminate flow occurs when the establish flow direction method cannot determine which direction commodities flow in a network.

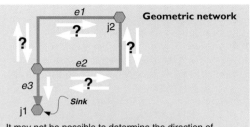

It may not be possible to determine the direction of flow given a configuration of sources, sinks, and enabled features. In this example, it is impossible to determine the flow through edges *e1* and *e2* because they form a cycle. Changing junction j2 to a source would break the cycle.

Logical network

Edge element table

Feature class	Feature ID	Sub-ID	Element ID	Flow direction
1	e1	1	10	Indeterminate
1	e2	1	11	Indeterminate
1	e3	1	12	With

The establish flow direction method will write "Indeterminate" as a flow direction when the flow direction cannot be established.

UNINITIALIZED FLOW

When a flow is isolated because the edges are disconnected from the rest of the network (that has flow), flow is said to be uninitialized.

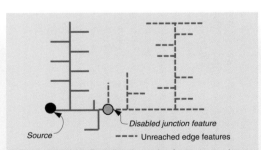

When establishing flow direction, edge features may be unreached because they are disconnected from the rest of the network. In this example, the unreached edges are disconnected because one of the junction features—a valve—is disabled.

WEIGHTS

Edges and junctions can have any number of weights associated with them. Weights are typically used to store the cost of traversing across an edge or through a junction. A typical weight is the length of the edge. Weights are created from field values on the edge and junction feature classes.

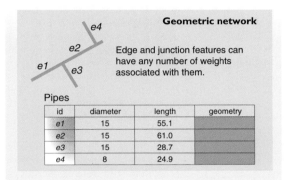

Geometric network

Edge and junction features can have any number of weights associated with them.

Pipes

id	diameter	length	geometry
e1	15	55.1	
e2	15	61.0	
e3	15	28.7	
e4	8	24.9	

Logical network

Edge element table

Feature class	Feature ID	Sub-ID	Element ID	Diameter	Length
1	e1	1	0	15	55.1
1	e2	1	1	15	61.0
1	e3	1	2	15	28.7
1	e4	1	3	8	24.9

Weights are stored with the logical network.

Weights are stored with the logical network so that analysis programs can access them efficiently. When a weight value is modified on a feature table, it is automatically updated in the logical network.

Any numeric field can be a weight. Determining which fields should be weights depends entirely on the types of analyses you wish to perform.

In ArcInfo, network analysis is a procedure that navigates through the connectivity of the network to yield some meaningful result, such as finding all elements upstream of a point or the shortest path between two points.

There are other ways to analyze networks, of course. For example, you could use the basic selection tools found in ArcMap to select edge features and then calculate statistics about them, such as the total edge length by type of edge. This is certainly a valid analysis on a network, but it is not a "network analysis" because the network connectivity is not involved.

Solvers

A program that performs network analysis is called a *solver,* because it solves a problem, such as isolating flow to an edge by turning off a set of valves. Inputs to this example flow isolation solver would be the logical network, the edge to isolate, and the set of junctions that are valves. The output would be the set of valves to turn off. There are no rules about the inputs and outputs of solvers, except that input always includes a logical network.

Solvers have user interfaces for specifying inputs and reporting outputs. Collections of solvers that perform similar tasks can usually be plugged into a common user interface framework. For example, the ArcInfo trace solvers are all accessed through a common toolbar. ArcMap is naturally part of the user interface for a solver, because ArcMap is where you graphically identify solver input, such as start points for a trace.

There are almost an infinite variety of solvers for the many types of network analyses. The ArcInfo strategy is to provide a rich suite of solvers that address the more common types of problems. For less common types of network analyses, developers can create solvers using any programming language that can access the ArcInfo components.

NetFlags

A *NetFlag* is a location on a network. Solvers use NetFlags to represent a multitude of real-world objects, such as stops for a shortest path, start points for tracing, locations of valves, locations of

services, and so on. NetFlags are not part of a logical network. They are used to describe any location in a network.

There are two kinds of NetFlags: *EdgeFlags* and *JunctionFlags*. NetFlag properties include the Logical Network element's feature class, feature ID, and feature sub-ID. An EdgeFlag additionally includes the percent along the edge element. This means that an EdgeFlag can fall anywhere along the edge, from zero percent (the from-junction) to 100 percent (the to-junction).

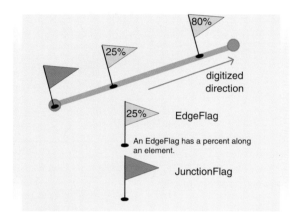

NetFlags are used to describe any location on a network. Examples include places to visit on a shortest path, the origin of a trace, a warehouse or a service center, or a valve, switch, or transformer. Solvers rely heavily on NetFlags to describe input parameters.

Barriers

Barriers are used by solvers to represent disabled logical network elements. Barriers do the same job as setting an element's enabled/disabled state to disabled, except that barriers are not stored with the logical network—they are known only to the solver. Barriers are just a way to temporarily disable elements. Barriers are either edge or junction elements.

There are four methods to capture and represent barriers to a solver. A well-designed solver will allow all four methods.

These are ways to set barriers to a solver:

- You can interactively add simple barriers.

- You can use the features in your selection set.

- You can disable feature classes.

- You can apply a weight as a filter.

TRACING

The ArcInfo network solvers currently work with a class of utility network problems termed *tracing*. Future versions of ArcInfo will include more solvers.

Tracing means to follow the flow in a network until some condition is met. When you hear problems expressed as "search against the flow until you find a transformer," or "follow the flow upstream to the first discharge point," or "trace upstream and find all valves," you are most certainly looking at using a trace solver to find the answer.

The ArcInfo trace solvers include upstream trace, downstream trace, isolation trace, and path trace.

WEIGHTS

Choosing which edge or junction attributes should become weights in the logical network depends on your collection of solvers. It is of no use to add a weight to a network if there are no solvers that can use it. For example, trace solvers typically do not use any weights—only the connectivity information found in the logical network.

For example, suppose you have a water distribution network with a numeric attribute containing the pipe manufacture ID. There is no need to add this attribute unless you have a solver that can use it. Even if you had a shortest path solver, it would not make sense to find the shortest path based on manufacturer ID.

But suppose you had a solver that could return all junctions that share edges of certain characteristics. In this case, you may want to use this solver to find all junctions where pipes from manufacture 100 connect with manufacture 151. In this case, it might make sense to add manufacturer ID as a weight.

Below is a table of just a few possible weight attributes and the types of solvers that would use these weights.

Weight description	Used for
Length of edge	Shortest path solvers. Many solvers have a need for length.
Diameter of pipe	Solvers that calculate pressure or head in a network.
Impedance (electrical resistance)	Calculating voltage drop in an electrical network.
Time to traverse an edge	Shortest path solvers.
Number of lanes on a street	Calculating traffic capacity or congestion on a street.
Road classification	Used to describe network hierarchy in hierarchical shortest path solvers.
Miles per hour	Used with a shortest path solver that allows dynamic calculation of weights.
Hazardous material route	Useful as a filter—find a path only on hazardous material routes.
Toll (cost to use a road)	Shortest path solvers based on actual cost.

Network solvers

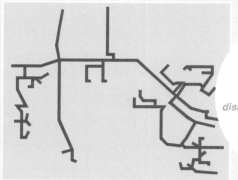

solver input

logical network

geometric network

simple barriers

weight filters

disabled feature classes

solver parameters

NetFlag collections

selection sets

In ArcMap, you can interactively set the state of a network through the placement of barriers, filters, and NetFlags by disabling certain feature classes in the geometric network, by creating a selection set, and by setting solver parameters.

solver

Trace Upstream ▼
Trace Upstream
Trace Downstream
Trace Connected
Trace Common Ancestors
Trace Loops
Trace Path

Once the state of the network is set, you specify the type of solver you want to execute.

solver output

edge and junction element collections

NetFlag collections

solver parameters

Solver output has several formats, such as a single number. Most typically, solvers output collections of elements and NetFlags.

A solver can display its trace result to a map through a solver renderer, which determines the set of network features that match the network elements in the trace. You can set several cartographic parameters to control the appearance of the trace.

solver renderer

draws features corresponding to network elements found from solver

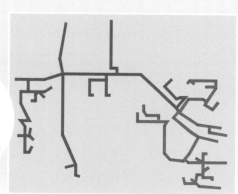

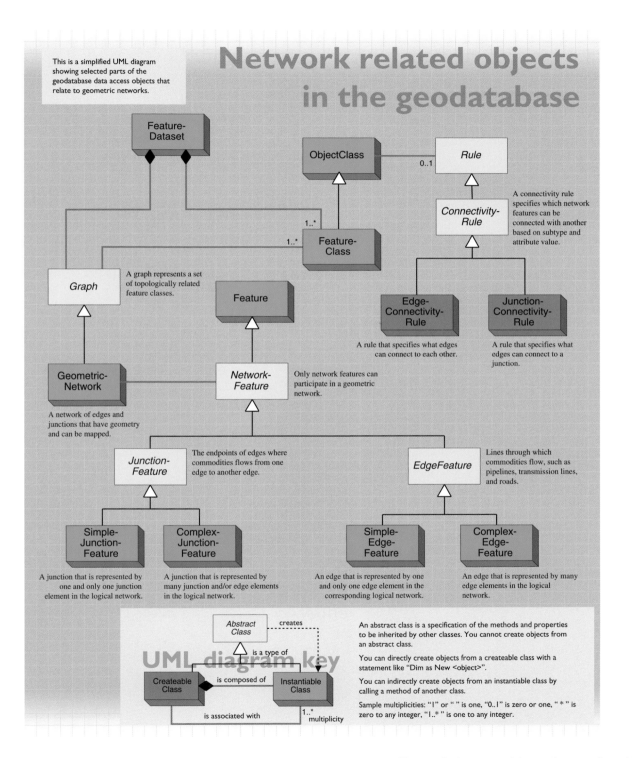

This is a simplified UML diagram showing selected parts of the geodatabase data access objects that relate to geometric networks.

Network related objects in the geodatabase

Feature-Dataset

ObjectClass

Rule

0..1

Connectivity-Rule

A connectivity rule specifies which network features can be connected with another based on subtype and attribute value.

1..*

1..*

Feature-Class

Graph

A graph represents a set of topologically related feature classes.

Feature

Edge-Connectivity-Rule

Junction-Connectivity-Rule

A rule that specifies what edges can connect to each other.

A rule that specifies what edges can connect to a junction.

Geometric-Network

Network-Feature

Only network features can participate in a geometric network.

A network of edges and junctions that have geometry and can be mapped.

Junction-Feature

The endpoints of edges where commodities flows from one edge to another edge.

EdgeFeature

Lines through which commodities flow, such as pipelines, transmission lines, and roads.

Simple-Junction-Feature

Complex-Junction-Feature

Simple-Edge-Feature

Complex-Edge-Feature

A junction that is represented by one and only one junction element in the logical network.

A junction that is represented by many junction and/or edge elements in the logical network.

An edge that is represented by one and only one edge element in the corresponding logical network.

An edge that is represented by many edge elements in the logical network.

UML diagram key

Abstract Class

creates

is a type of

Createable Class

is composed of

Instantiable Class

is associated with

1..*
multiplicity

An abstract class is a specification of the methods and properties to be inherited by other classes. You cannot create objects from an abstract class.

You can directly create objects from a createable class with a statement like "Dim as New <object>".

You can indirectly create objects from an instantiable class by calling a method of another class.

Sample multiplicities: "1" or " " is one, "0..1" is zero or one, " * " is zero to any integer, "1..* " is one to any integer.

Network object model

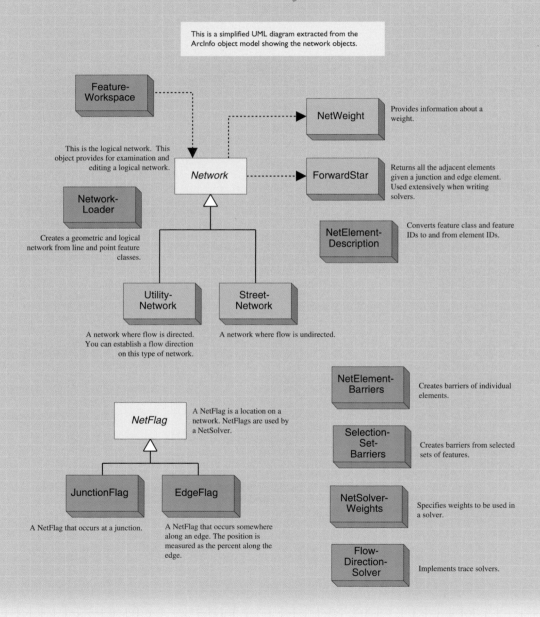

This is a simplified UML diagram extracted from the ArcInfo object model showing the network objects.

Feature-Workspace

NetWeight — Provides information about a weight.

This is the logical network. This object provides for examination and editing a logical network.

Network

ForwardStar — Returns all the adjacent elements given a junction and edge element. Used extensively when writing solvers.

Network-Loader

Creates a geometric and logical network from line and point feature classes.

NetElement-Description — Converts feature class and feature IDs to and from element IDs.

Utility-Network

Street-Network

A network where flow is directed. You can establish a flow direction on this type of network.

A network where flow is undirected.

NetElement-Barriers — Creates barriers of individual elements.

NetFlag — A NetFlag is a location on a network. NetFlags are used by a NetSolver.

Selection-Set-Barriers — Creates barriers from selected sets of features.

JunctionFlag

EdgeFlag

NetSolver-Weights — Specifies weights to be used in a solver.

A NetFlag that occurs at a junction.

A NetFlag that occurs somewhere along an edge. The position is measured as the percent along the edge.

Flow-Direction-Solver — Implements trace solvers.

9

Cell-based modeling with rasters

Some geographic phenomena can be represented best as locations on a grid with values. This data structure is called a raster. In this chapter:

- Representing geography with rasters

- Using raster data

- Raster data model

- Raster display and analysis

- The spatial context of rasters

- Raster formats

- Raster object model

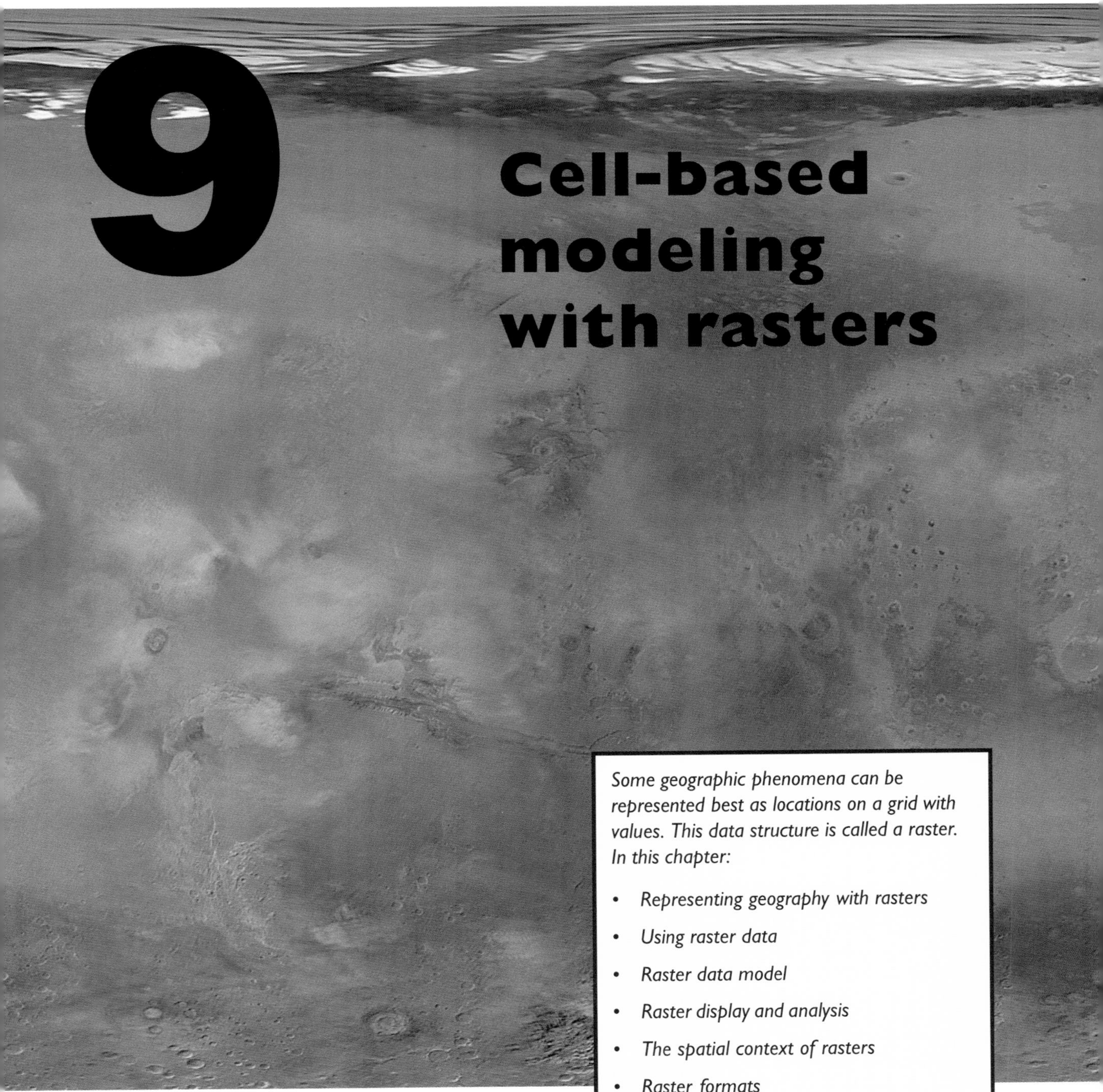

Mosaicked image of Mars on a simple cylindrical projection. Mars Global Surveyor, Malin Space Science Systems and JPL/NASA, 1999.

A raster is a rectangular array of equally spaced cells, which taken as a whole represent thematic, spectral, or picture data. Raster data can represent everything from qualities of a land surface such as elevation or vegetation, to satellite images, scanned maps, and photographs.

The raster data format is very simple but supports a rich variety of data types. This chapter discusses the ways that geographic data is represented in rasters.

FORMS OF RASTER DATA

Raster data can be entered into a GIS through imaging systems or calculated from other data.

Satellite imagery

Acquiring satellite images is a cost-effective way to map a sizeable part of the world at small to moderate map scales.

Satellite images are perhaps the best system in a GIS to capture temporal changes in the landscape. You can compare and analyze scenes of the same area from different seasons or years.

Images can show a scene in color or black and white. Color information is stored as an RGB composite pixel value or as a set of raster bands representing several or many colors.

Some satellite imagery services allow you to order fresh images, only days or hours old. These are an important asset for managing environmental events such as floods or forest fires.

Aerial imagery

To create detailed maps, airplanes with special large-format photographic and digital cameras record strips of images that overlap to cover an area.

These images are then rectified for scale distortion caused by the surface's shape.

Scanned maps

Sometimes, the best basemap is a published map that is scanned. These can be assigned a geographic reference, so that the scanned map can be precisely registered with other geographic data.

The USGS (United States Geological Survey) quadrangle maps are a good example of maps that are scanned. These maps show terrain, place names, rivers, roads, and major features.

Pictures

Besides images of land, rasters are also used for photographs of features. Photographs of features can be an important augmentation to the information presented in a map.

Converted data

Rasters can also be generated from other data sources, such as feature datasets and TIN datasets. Analysis results done from raster surface studies can be used to make a slope map. An image classification can be used to make a land-cover map.

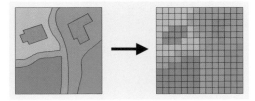

TYPES OF RASTER DATA

There are two general categories of raster data: thematic data and image data. Thematic data can be used for geographic analysis of land use. Image data is used for basemaps to other geographic data and to derive thematic data.

Thematic data

The value of each cell (or pixel) in a raster can be a measured quantity or classification. When drawn, these rasters are thematic maps.

Spatially continuous data

The values of raster cells can represent a measured quantity such as elevation, pollution concentration, or rainfall. The value from one cell to another varies slightly, and collectively, these values can model some type of surface.

Cell values for spatially continuous data represent a sampled quantity at cell centers.

Spatially discrete data

Values for raster cells can represent a category or classification of data, such as land ownership type or vegetation type. The value from one cell to another is typically either identical or changes abruptly. This type of data appears as a set of zonal regions with common values, such as land-use maps or forest stands.

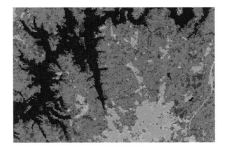

Cell values for spatially discrete data represent a classification that applies to the full area of a cell.

Image data

The preponderance of raster data is captured by imaging systems mounted on satellites and airplanes.

Spectral and picture data

Imaging systems record rasters based on light reflectance values at one or many bands of the electromagnetic spectrum.

Picture data usually captures the red, green, and blue portions of the spectrum for display on a monitor or map, but certain satellite images capture many bands that are used to analyze surface geology and vegetation.

Rasters are used for display and analysis. These topics illustrate just a few of the uses of rasters.

Basemap

Often, rasters are used as a visual backdrop on a map. They are the bottom layer on which vector and surface data is drawn.

For example, you can draw road features on top of a raster image of a city and instantly see what parts of the road network need updating.

Using a good raster image as a basemap layer adds depth to the map and enhances the user's trust in the map. A raster basemap is a good check on most other geographic data.

Land-use scenario

Raster data is ideal for modeling and mapping land use and land-use change. Most land-use studies begin with satellite or aerial imagery that is interpreted and then categorized into classes such as urban, agriculture, and deciduous forest.

Over time, these studies are repeated, and differences between years can be analyzed.

Hydrological analysis

Terrain information is commonly available in a raster form with elevations for cell values. This is called a digital elevation model (DEM).

Raster GIS tools let you to determine the direction of water flow across the landscape, downstream accumulation of precipitation, and the delineation of drainage basins or watersheds.

This model is the basis for doing hydrologic analysis such as runoff prediction from a storm and which structures are at risk of flooding. This information is useful for mapping floodplains and for determining flood insurance rates.

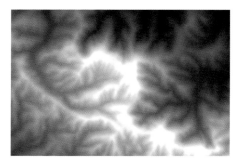

Environmental analysis

Because data such as land cover, vegetation type, and terrain are commonly stored as rasters, most environmental analysis involves raster data.

Raster GIS analysis tools have evolved to use such data to solve problems at many scales. These range from continental forest succession as a result of global warming to local wildlife habitat changes resulting from urbanization.

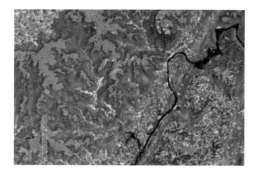

Terrain analysis

Digital elevation models contain elevations for cell values. ArcInfo contains raster analysis tools to study the visibility, slope, aspect, and curvature calculations that are often used as part of a large study, such as land-use planning or site selection.

You can pose questions such as, "Find all the locations between 2,500 and 5,000 feet in elevation, facing south or southeast, on a slope of less than 12 percent, with 3 miles of visibility."

ArcInfo also includes display functions for digital elevation models, an example of which is analytic hillshading, which produces a realistic view of terrain. These are some maps that show surface display methods.

This map shows elevation by shaded colors. The green cells show lower elevation. The red, pink, and white cells show higher elevation.

For each cell in a digital elevation model, a hillshade map draws shades that simulate the illumination of a surface, based on the angle between the sun and the slope of the local surface.

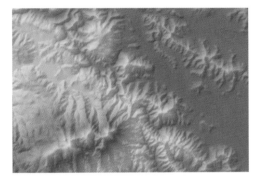

This map shows elevation combined with hillshading. The combination of these display methods creates an attractive map that simultaneously shows heights and the surface shape.

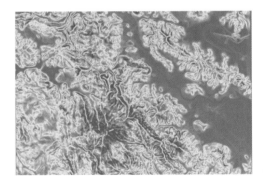

This map shows the slope of a terrain. The red cells show steep areas and the green cells show flat areas.

Rasters are made of cells. A cell is a uniform unit that represents a defined area of the earth, such as a square meter or square mile.

Each raster cell has a value that represents a spectral reflectance or a characteristic at that position, such as a soil type, census tract, or vegetation class. Additional values of the cell can be stored in an attribute table.

The size chosen for a grid cell of a study area depends upon the data resolution required for the most detailed analysis. The cell must be small enough to capture the required detail, but large enough so that computer storage and analysis can be performed efficiently. The more homogeneous an area is for critical variables such as topography and land use, the larger the cell size that can be used with accuracy.

CELL ATTRIBUTES

The value associated with a cell defines the class, group, category, or measure at the cell position. Cell values are numbers: integer or floating point.

Cell locations with the same value belong to the same zone. Cells of the same zone do not have to be connected.

When an integer value is used for cells, it may be a code for a much more complex identification. For example, the value four may equate to single-family residential parcels on a land-use grid. Associated with the value of four might be a series of attributes, such as the average commercial value, average number of inhabitants, or census code.

There is usually a one-to-many relationship between the grid cell values (or codes) and the number of cells that are assigned the code. That is, there might be 400 cells with the value four (representing single-family residential) and 150 cells associated with the value five (representing commercial zoning) on the land-use grid.

The code value occurs many times in the raster, but only once in the attribute table, which stores additional attributes for the code. This design reduces storage and simplifies updating. A single change to an attribute can be applied to several hundred instances of that value.

TYPES OF DATA

Each cell in a raster has one value. The cell values in a raster can represent one of the following four general types of data.

Nominal data

A value of nominal data identifies one entity from another. These values establish the group, class, member, or category with which the geographic entity at the position of the cell is associated. These values are qualities, not quantities, with no relation to a fixed point or a linear scale. Coding schemes for land use, soil types, or any other attribute qualify as a nominal measurement.

Ordinal data

A value of ordinal data determines the rank of an entity versus other entities. These measurements show place, such as first, second, or third, but they do not establish magnitude or relative proportions. You cannot infer a quantitative difference, such as how much an entity is larger, higher, or denser than the others.

Interval data

A value of interval data represents a measurement on a scale such as time of day, temperature in Fahrenheit degrees, and pH value. These values are on a calibrated scale but are not relative to a true zero point. You can make relative comparisons between interval data, but their measure is not meaningful when compared to the zero point of the scale.

Ratio data

A value of ratio data represents a measure on a scale with a fixed and meaningful zero point. Mathematical operations can be used on these values with predictable and meaningful results. Examples of ratio measurements are age, distance, weight, and volume.

Inside a raster

Rasters are two-dimensional arrays of cells (or pixels). The height and width of each cell are fixed and the same. A raster spans a rectangular area.

Each cell has a value. This value can represent many qualities of a location, including reflectance, color, precipitation, and elevation.

Rasters have an integer coordinate space. You can determine the coordinate of a cell by counting columns from the left and rows from the top. Row and column values begin with 0.

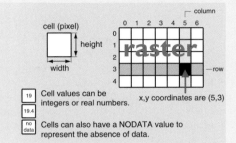

Cell values can be integers or real numbers.

Cells can also have a NODATA value to represent the absence of data.

x,y coordinates are (5,3)

The attribute table

Value	Count	Type	Code
23	7	Fir	400
29	18	Juniper	410
31	10	Aspen	420
37	18	Piñon	500
41	4	Cottonwood	510
43	7	Walnut	600

Rasters that have integer valued cells can be defined with an optional attribute table, which records attributes for each unique cell value.

You can add custom fields to the attribute table.

Types of data represented in cells

The data stored in a raster can be categorized as one of these types.

Nominal data

Fir / Juniper / Aspen / Piñon / Cottonwood / Walnut

Nominal data values are categorized and have names. The data value is an arbitrary type code. Examples are soil types and land use.

21	17	17	18	22	18
18	16	17	19	24	19
21	19	19	19	22	22
26	23	21	20	18	21
24	23	18	16	20	19
18	14	16	17	19	20

Nominal and ordinal data represent discrete categories. They are best represented with integer cell values.

Ordinal data

very good / good / moderate / poor

Ordinal data values are categorized, have names, and the value is in a numerical rank. Examples are land suitability classifications and soil drainage rank.

Interval data

700–709 / 710–719 / 720–729 / 730–739 / 740–749 / 750–759

Interval data values are numerically ordered and the interval difference is meaningful. Examples are voltage potential and difference in concentration.

21.1	17.3	17.2	18.1
18.5	16.2	17.3	19.1
21.0	19.1	19.4	19.2
26.3	23.1	21.6	20.5

Interval and ratio data present continuous phenomena and are usually measured with real cell values.

Ratio data

0.0–10.0 / 10.1–20.0 / 20.1–30.0 / 30.1–40.0 / 40.1–50.0

Ratio data values measure a continuous phenomenon with a natural zero point. Examples are rainfall and population.

The rendition of rasters

A raster can have one or many bands. The cell values of rasters can be drawn in a variety of ways. These are some of the ways to display rasters by cell values.

Displaying single-band rasters

Cell values in single-band rasters can be drawn in these three basic ways.

Monochrome image

0	0	0	0	1	1
1	0	0	1	1	0
1	0	1	1	0	0
0	0	0	1	1	0
1	1	0	0	0	1
0	1	1	1	0	0

0	1

In a monochrome image, each cell has a value of 0 or 1. They are often used for scanning maps with simple linework, such as parcel maps.

Grayscale image

68	124	0	170	86	0
234	187	68	251	10	236
76	124	218	132	201	66
124	16	118	183	32	255
126	191	198	251	141	56
41	255	243	162	212	152

0	255

In a grayscale image, each cell has a value from 0 to 255. They are often used for black-and-white aerial photographs.

Display colormap image

1	5	3	2	2	4
5	2	4	2	5	1
5	5	5	5	3	3
2	1	2	4	1	3
4	4	4	1	1	3
2	4	2	1	3	3

Colormap

		red	green	blue
1	◄►	255	255	0
2	◄►	64	0	128
3	◄►	255	32	32
4	◄►	128	255	128
5	◄►	0	0	255

One way to represent colors on an image is with a colormap. A set of values is arbitrarily coded to match a defined set of red-green-blue values.

Displaying multiband rasters

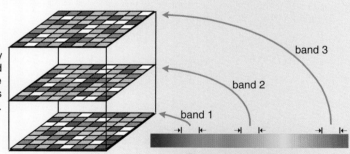

Raster datasets have one or many bands. In multiband rasters, a band represents a segment of the electromagnetic spectrum that has been collected by a sensor.

Electromagnetic spectrum

Bands often represent a portion of the electromagnetic spectrum, including ranges not visible to the eye—the infrared or ultraviolet sections of the spectrum.

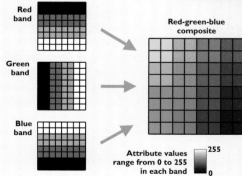

Red band

Green band

Blue band

Red-green-blue composite

Attribute values range from 0 to 255 in each band

Multiband rasters are often displayed as red-green-blue composites. This band configuration is common because these bands can be directly displayed on computer displays, which employ a red-green-blue color rendition model.

Calculations with rasters

Raster operators

When you are studying an area, you may want to apply a suitability analysis. To do this, you would select rasters with values such as rainfall, soil alkalinity, and insolation and apply a series of operators according to your formula for suitability. Operators can be arithmetic, Boolean, relational, bitwise, combinatorial, logical, accumulative, and assignment.

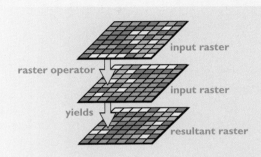

input raster

raster operator

input raster

yields

resultant raster

Map calculation

Mathematical operations can be applied to two rasters and the result is in the output raster. Functions include +, −, /, *, Log, Exp, Sin, Cos, and Sqrt.

4	2
1	3

+

3	4
1	1

=

7	6
2	4

Map query

You can apply Boolean and logical operators on two rasters to create an output raster with true/false values. Operators include And, Or, XOr, Not, >, >=, =, <>, <, and <=.

sand	clay
clay	sand

+

dry	dry
wet	wet

=

true	false
false	false

Raster functions

There are many raster functions. Each can accept one or many rasters as input and generate one or several rasters with the calculated results.

Local function

Local functions perform a calculation on a single cell at a time. The neighboring cells do not influence the result. The functions can be applied to one raster or several overlaid rasters. Local functions include trigonometric, exponential, logarithmic, reclassification, selection, and statistical functions.

Focal function

Focal functions perform a calculation on a single cell and its neighboring cells. A neighborhood can be a rectangle, circle, annulus (doughnut), or wedge. These functions can return the mean, standard deviation, sum, or range of values within the immediate or extended neighborhoods.

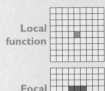

Zonal function

Zonal functions perform a calculation on a zone, which is a set of cells with a common value. The cells that form a zone can be discontinuous. There are two categories of zonal functions: statistical and geometric. These functions include area, centroid, perimeter, ranges, and sum calculations.

Global function

Global functions perform a computation on the raster as a whole. Examples are the calculation of Euclidean distances, weighted cost distances, and watershed delineation.

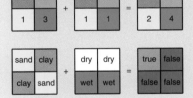

Input raster(s)

applied to a raster function yields

resultant raster(s).

When a raster is captured with devices like satellite imaging systems or desktop scanners, the raw data is just rows and columns of cells. In order to use such data in a GIS either to draw on the screen with other data or overlay in an analysis operation, the data must be in a common coordinate system. This is a real-world coordinate system.

The rows and columns of a raster are always parallel to the x- and y-axes of the coordinate system.

Georeferencing

Georeferencing is the process of establishing a relationship between the raster's (row, column) coordinate system, sometimes called image space, and a real-world (x,y) coordinate system, called map space. A similar transformation process occurs when establishing the relationship of feature data in one coordinate system (i.e., digitizer units) to another map coordinate system.

You can define a transformation to register the raster to real-world coordinates. This lets your raster in the same space as your other geographic data, such as vector features or a surface in a TIN dataset.

To georeference a raster, it needs to be registered to a coverage, map, or set of coordinates that are in map space. The registration process is normally an interactive one where common locations are selected both in the raster and the other geographic dataset. For imagery, these are normally things like road intersections that are easily identifiable in both datasets. Once the common locations are selected, a polynomial transformation is built to model the scale, rotation, and skew between the two coordinate systems.

The georeferencing information is stored internally to some raster formats, such as the ESRI ARC GRID™ or ERDAS IMAGINE®, or in external files such as the raster auxilliary file (.aux) or the world file for other formats such as JPEG or BMP.

Using this information, the raster can be transformed on-the-fly and drawn in the map space of your other data. If you have also stored the map projection information, it can also be projected into other coordinate systems.

Rectification

To align an image axis with a map-space axis, an image must be rectified by resampling it based on the transformation built during the registration process. In resampling, a mesh is overlaid on the raster and a value is assigned to each cell according to the center's proximity to the values of the centers of the cells in the rotated raster. The values assigned to the output raster will be determined by the type of resampling: nearest neighbor assignment, bilinear interpolation, or cubic convolution.

You may wish to rectify a raster to remove skew or rotation, to orient its cells orthogonally to the map orientation. The primary reason not to rectify is because any such resampling of a raster will induce a small amount of error. This amount of error is not something you can see but can be important in multispectral analysis where minute differences in cell values can be significant.

Raster pyramids

Large rasters can display quickly if pyramids have been created for them. They often contain more information than can be displayed on the screen. If pyramids are not present, then the entire raster dataset must be investigated, and many calculations must be made to choose which subset of data cells is sent to the display. Pyramids are a way to store reduced-resolution copies of the raster, and by choosing a resolution that is similar to the amount of display area, there are fewer cells to investigate and fewer calculations, which therefore decreases the display time.

Referencing cell positions

Rasters are stored as arrays of cells (pixels) and can be displayed on the map's coordinate system. Rasters of geographic areas have a display transformation that converts cell units to map coordinates.

Image to world affine transformation

$x' = Ax + By + C$
$y' = Dx + Ey + F$, where

x is column count in image space.
y is row count in image space.
x' is horizontal value in coordinate space.
y' is vertical value in coordinate space.

A is width of cell in map units.
B is a rotation term.
C is the x' value of the center of upper-right cell.
D is a rotation term.
E is negative of height of cell in map units.
F is the y' value of the center of upper-right cell.

Six parameters define how a raster's rows and columns transform onto map coordinates.

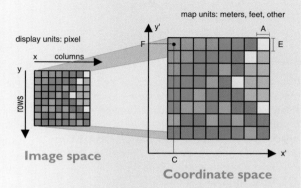

map units: meters, feet, other

display units: pixel

x   columns

y

rows

Image space

Coordinate space

Value applies to center point of cell

For certain types of data, the cell value represents a measured value at the center point of the cell. An example is a raster of elevation values.

Value applies to whole area of cell

For most data, the cell value represents a sampling of a phenomenon and the value is presumed to represent the whole cell square.

When you add raster datasets in ArcInfo, you have the option to create associated pyramids. They comprise a set of rasters that are progressively downsampled by a factor of two.

As you zoom out and the raster cells grow smaller than the screen display pixels, ArcMap will select one of the pyramided rasters to draw.

The purpose of pyramids is to optimize display performance.

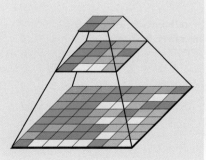

You can display most industry-standard raster formats in ArcInfo. This is a summary of those formats with a description of their attributes.

- ADRG (ARC Digitized Raster Graphics) data consists of raster images scanned and distributed on CD–ROM by the United States National Imagery and Mapping Agency (US NIMA). ADRG is geographically referenced using the equal arc-second raster chart/map (ARC) system in which the globe is divided into 18 latitudinal zones.

- CADRG (Compressed ARC Digitized Raster Graphics) data is the same as ADRG data, but is compressed with a nominal ratio of 55:1.

- CIB (Controlled Image Base) images are panchromatic (grayscale) images that have been georeferenced and corrected for distortion due to topographic relief. They are distributed by US NIMA and are commonly used as basemaps.

- DTED Level 1 & 2 is elevation data usually produced by US NIMA.

- ERDAS® 7.5 GIS images are single-band thematic images produced by the ERDAS 7.5 (or earlier) image-processing software.

- ERDAS 7.5 LAN images are single- or multiband continuous images produced by the ERDAS 7.5 (or earlier) image-processing software.

- ERDAS raw data is used to read and display uncompressed, band interleaved by line, band interleaved by pixel, and band sequential image data. Through an ASCII file that describes the layout of the image data, black-and-white, grayscale, pseudocolor, and multiband image data can be displayed without translation into a proprietary format.

- ERDAS IMAGINE files are produced using IMAGINE image-processing software. IMAGINE files can store both continuous and discrete single-band and multiband data.

- ER Mapper files are produced using the ER Mapper image-processing software.

- ESRI BIL/BIP/BSQ data is used to read and display uncompressed, band interleaved by line, band interleaved by pixel, and band sequential image data. Through an ASCII file that describes the layout of the image data, black-and-white, grayscale, pseudocolor, and multiband image data can be displayed without translation into a proprietary format.

- ESRI ARC GRID data supports 32-bit integer and 32-bit floating-point raster grids. Grids are especially suited for both discrete and continuous phenomenoma and for performing spatial modeling and analysis of flows, trends, and surfaces such as hydrology.

- ESRI ARC GRID Stacks and Stack files are used to reference multiple ESRI grids as a multiband raster data set. A stack is stored in a directory structure similar to a grid or coverage. A stack file is a simple text file that stores the path and name of each ESRI grid contained within it on a separate line.

- GIF (Graphics Interchange File) is CompuServe's standard for defining generalized color raster images. This format allows high-quality, high-resolution graphics to be displayed on a variety of graphics hardware and is intended as an exchange and display mechanism for graphics images.

- JFIF (JPEG File Interchange Format) applies the JPEG compression technique for storing full-color and grayscale images.

- MrSID™ (Multiresolution Seamless Image Database) is a multiresolution wavelet-based image format with a high compression ratio. It allows fast access of large amounts of data at any scale.

- TIFF (Tagged Image File Format) is widely used in desktop publishing and serves as an interface to several scanners and graphic arts packages. TIFF supports black-and-white, grayscale, pseudo-color, and true color images, all of which can be stored in compressed or uncompressed format.

- Microsoft Windows/IBM® OS/2® Bitmap (BMP) images are usually used to store pictures or clip art. These images can be moved between different applications on Windows or OS/2 platforms.

Raster format	Supported data types	Supports multiband	Supported compressions	Supports colormaps	File structure
ADRG, ARC Digitized Raster Graphics—image, overview, and legend	8-bit unsigned integer	yes, always 3 bands	none	no	multiple files with .img, .ovr, and other extensions
CADRG Compressed ARC Digitized Raster Graphics	8-bit unsigned integer	yes, always 3 bands	vector quantization	no	single file, no standard extension
CIB Controlled Image Base	8-bit unsigned integer	no	vector quantization	no	single file, no standard extension
DTED Level 1 & 2 Digital Terrain Elevation Data	16-bit signed integer	no	none	no	single file, various extensions
ERDAS 7.5 GIS	1-, 2-, 4-, 8-, 16-bit unsigned integer	no	none	yes	multiple files, data file with .gis extension, colormap with .trl extension
ERDAS 7.5 LAN	8-, 16-bit unsigned integer	yes	none	no	multiple files, data file with .lan extension, statistics file with .sta extension
ERDAS Raw	1-, 2-, 4-, 8-, 16-, 32-bit unsigned integer	yes	none	no	multiple files, header file with .raw extension, data file usually same name as header without extension
	16-, 32-bit signed integer	yes	none	no	
	32-, 64-bit floating point	yes	none	no	
ERDAS IMAGINE	1-, 2-, 4-, 8-, 16-, 32-bit unsigned integer	yes	none, adaptive run length compressed	yes	single file with .img extension
	8-, 16-, 32-bit signed integer	yes	none, adaptive run length compressed	yes	
	32-, 64-bit floating point	yes	none	yes	
	64-, 128-bit complex	yes	none	yes	
ER Mapper	8-, 16-bit unsigned integer	yes	none	yes	multiple files, header file with .ers extension, data file is usually the same name as header file without extension
	32-bit unsigned integer	yes	none	no	
	8-, 16-bit signed integer	yes	none	yes	
	32-bit signed integer	yes	none	no	
	32-, 64-bit floating point	yes	none	no	
ESRI BIL Band interleaved by line	1-, 4-, 8-, 16-, 32-bit unsigned integer	yes	none	yes	multiple files, data file has .bil, .bip, or .bsq extension, header file has .hdr extension, colormap file has .clr extension, statistics file has .stx extension
ESRI BIP Band interleaved by pixel	1-, 4-, 8-, 16-, 32-bit unsigned integer	yes	none	yes	
ESRI BSQ Band sequential	1-, 4-, 8-, 16-, 32-bit unsigned integer	yes	none	yes	
ESRI ARC GRID	32-bit signed integer	no	adaptive run length compressed	yes	ARC GRID and ARC GRID Stack is a folder of files with .adf extension and colormap with .clr extension. ARC GRID Stack file has a possible .stx extension
	32-bit floating point	no	none	no	
ESRI ARC GRID Stack	32-bit signed integer	always	adaptive run length compressed	no	
	32-bit floating point	always	none	no	
ESRI ARC GRID Stack File	32-bit signed integer	always	adaptive run length compressed	no	
	32-bit floating point	always	none	no	
GIF Graphics Interchange File	8-bit unsigned integer	no	LZW	yes	single file with .gif extension
JFIF (JPEG) JPEG File Interchange Format	8-bit unsigned integer	yes, 1 or 3 bands	JPEG	no	single file with .jpg, .jpeg, or .jfif extension
MrSID Multiresolution Seamless Image Database	8-bit unsigned integer	yes, 1 or 3 bands	wavelet	no	single file with .sid extension
TIFF Tagged Image File Format (GeoTIFF tags are supported)	1-bit unsigned integer	no	none, CCITT Group 3 1-D, CCITT Group 4, PackBits, LZW	yes	single file with .tif, .tiff, or .tff extension
	4-bit unsigned integer	no	none, PackBits, LZW	yes	
	8-bit unsigned integer	yes	none, PackBits, LZW, JPEG	yes	
	16-bit unsigned integer	yes	none	yes	
	32-bit unsigned integer	no	none	yes	
	8-, 16-, 32-bit signed integer	no	none	no	
	32-, 64-bit floating point	no	none	no	
Windows/OS/2 Bitmap BMP or Microsoft Windows/IBM OS/2 Bitmap or Device-Independent Bitmap (DIB)	1-bit unsigned integer	no	none	yes	single file with .bmp extension
	4-bit unsigned integer	no	none, run length encoded	yes	
	8-bit unsigned integer	1 or 3 bands	none, run length encoded (single band)	yes	

Raster objects

This is a simplified UML diagram extracted from the ArcInfo Object Model diagram that shows the raster data access objects.

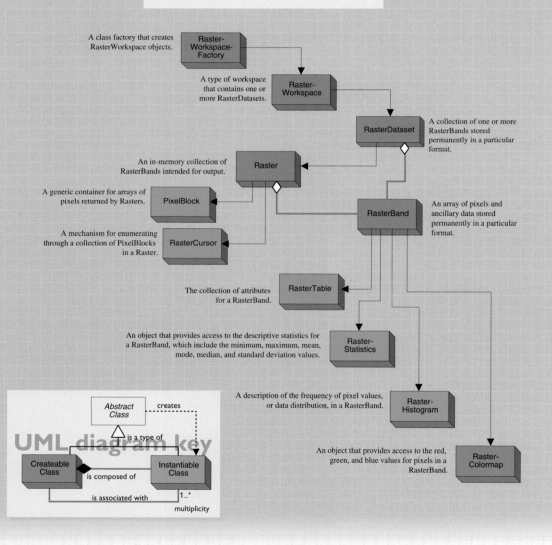

A class factory that creates RasterWorkspace objects.

Raster-Workspace-Factory

A type of workspace that contains one or more RasterDatasets.

Raster-Workspace

RasterDataset

A collection of one or more RasterBands stored permanently in a particular format.

An in-memory collection of RasterBands intended for output.

Raster

A generic container for arrays of pixels returned by Rasters.

PixelBlock

RasterBand

An array of pixels and ancillary data stored permanently in a particular format.

A mechanism for enumerating through a collection of PixelBlocks in a Raster.

RasterCursor

The collection of attributes for a RasterBand.

RasterTable

An object that provides access to the descriptive statistics for a RasterBand, which include the minimum, maximum, mean, mode, median, and standard deviation values.

Raster-Statistics

A description of the frequency of pixel values, or data distribution, in a RasterBand.

Raster-Histogram

An object that provides access to the red, green, and blue values for pixels in a RasterBand.

Raster-Colormap

UML diagram key

Abstract Class

creates

is a type of

Createable Class

is composed of

Instantiable Class

is associated with

1..*

multiplicity

10 Surface modeling with TINs

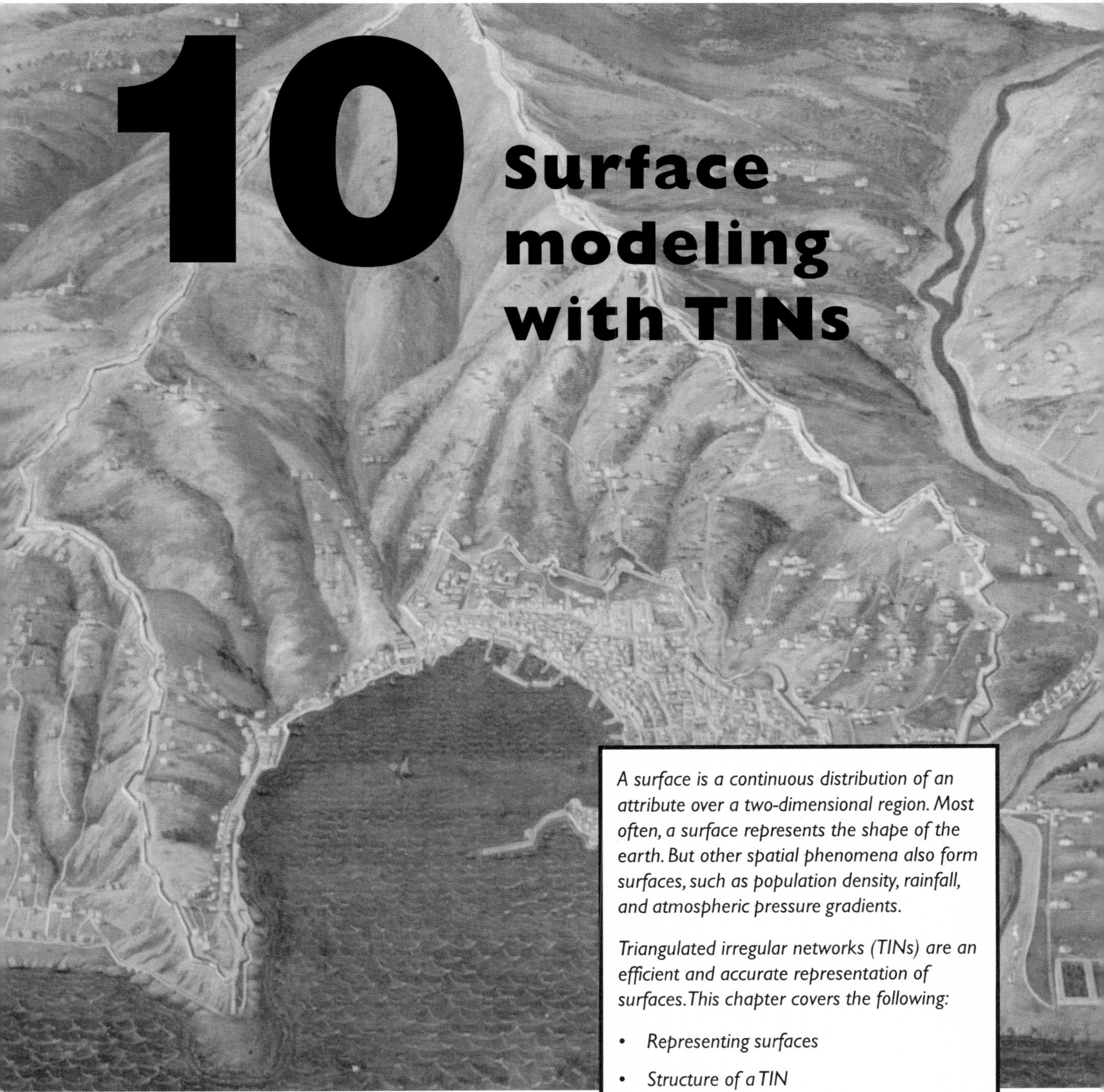

Genua, The Gallery of Maps in the Vatican, 1636.

A surface is a continuous distribution of an attribute over a two-dimensional region. Most often, a surface represents the shape of the earth. But other spatial phenomena also form surfaces, such as population density, rainfall, and atmospheric pressure gradients.

Triangulated irregular networks (TINs) are an efficient and accurate representation of surfaces. This chapter covers the following:

• Representing surfaces

• Structure of a TIN

• Modeling surface features

Most of the geographic objects on a map lie on the surface of the earth.

Entities such as buildings, roads, and wells are usually modeled as features—two-dimensional vector shapes with attributes, relationships, and behavior in feature datasets inside a geodatabase.

Other entities, such as drainages, ridges, and peaks, are integral components of a surface. You can represent these entities as features—their shapes can be neatly drawn on a map. But if you want to perform some form of surface analysis such as hydrography or viewshed studies, you must embed these discrete entities within a continuous representation of a surface.

The previous chapter discussed the versatile use of raster datasets to represent a wide range of phenomena, including surfaces. This chapter first compares the utility of rasters and triangulated irregular networks (TINs) for surface modeling, and then further examines the TIN data representation.

QUALITIES OF SURFACES

Surfaces represent a continuous field of z-values with an infinite number of points. Computers and the notion of infinity are generally incompatible—some type of sampling is necessary to derive an acceptable approximation of a surface in a GIS.

ArcInfo uses two representations to model surfaces: rasters and TINs. Rasters represent a surface as a regular grid of locations with sampled or interpolated z-values. TINs represent a surface as a set of irregularly located points that form a network of triangles with z-values at each node.

Both raster and TIN representations have merit for surface modeling; the context of available source data, and the scope of analysis and cartography to be supported, will guide which representation is better for a particular application.

The raster representation of a surface

Rasters represent surfaces as a regular grid of uniformly spaced locations with z-values. You can estimate a surface value for any location by interpolating z-values among immediate grid points.

The resolution of the grid—the width and height of cells—determines the precision of the raster representation.

Rasters are the most common representation of surfaces because elevation data is widely available in this form at low cost. An example of raster surface data are the digital elevation models (DEMs) produced by the United States Geological Survey.

Rasters support a rich set of spatial analysis such as spatial coincidence, proximity, dispersion, and least-cost paths, which can be performed rapidly.

The disadvantages of the raster representation is that surface discontinuities such as ridges are not well represented and precise locations for features such as peaks are lost in the grid sampling of rasters.

Rasters are appropriate for small-scale mapping applications where positional accuracy is not paramount and where surface features do not need to be characterized exactly.

The TIN representation of a surface

TINs represent surfaces as contiguous nonoverlapping triangular faces. You can estimate a surface value for any location by simple or polynomial interpolation of elevations in a triangle.

Because elevations are irregularly sampled in a TIN, you can apply a variable point density to areas where the terrain changes sharply, yielding an efficient and accurate surface model.

A TIN preserves the precise location and shape of surface features. Areal features such as lakes and islands are represented by a closed set of triangle edges. Linear features such as ridges are represented by a connected set of triangle edges. Mountain peaks are represented by a triangle node.

TINs support a variety of surface analyses such as calculating elevation, slope, and aspect, performing volume calculations, and creating profiles on alignments. The disadvantage of TINs is that they are often not readily available and require data collection.

TINs are well suited for large-scale mapping applications where positional accuracy and shapes of surface features are important.

Comparing rasters and TINs for representing surfaces

Surfaces can be modeled with rasters or TINs. Each model has advantages and limitations.

Rasters are a simpler model of a surface. Digital terrain data is widely accessible in raster format.

TINs can produce a more accurate representation of surfaces and features, but usually require a data collection effort.

Raster representation of a surface

```
+451   +454   +457   +459   +458

+453   +455   +456   +461   +461

+454   +459   +458   +465   +467

+456   +460   +462   +473   +469

+458   +462   +464   +469   +465
```

TIN representation of a surface

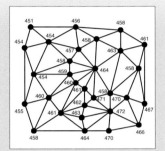

	Raster	TIN
Accuracy of surface model	The precision of a raster surface model is determined by the cell dimensions. To increase the accuracy of a raster surface model, the entire raster must be resampled at a higher resolution.	A TIN surface model has a variable point density that varies on the degree of change of slope. To make a TIN more accurate, additional mass points, breaklines, and polygons can be added.
Fidelity of surface features	Rasters sample the z-values of surface features on a regular grid. Features such as peaks and ridges cannot be located to a position more accurate than the grid resolution.	TINs are designed to capture and represent surface features such as streams, ridges, and peaks. These features are stored with precise coordinates, and slope discontinuities such as ridges are modeled with breaklines.
Surface analysis	Spatial coincidence Proximity Dispersion Least-cost path	Elevation, slope, aspect calculations Contour derivation from surface Volume calculations Vertical profiles on alignments Line-of-sight analysis
Sample applications	Small-scale surface display and modeling Modeling of pollutant dispersion Identification of watershed basins Hydrologic analysis of flood zones	Volumetric calculations for roadway design Drainage studies for land development Generation of high-quality elevation contours Perspective displays of buildings on a landscape

The TIN data structure lets you accurately represent any type of surface. Not only can elevations be interpolated for any location within a TIN, but natural features that form breaks in a surface's slope, such as ridges and streams, can also be stored.

DEFINITION OF A TIN

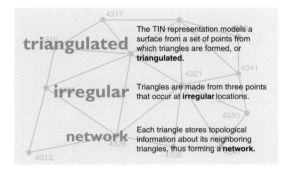

The term *triangulated irregular network* is a concise description of the characteristics of a TIN.

"Triangulated" refers to the forming of an optimized set of triangles from a set of points. Triangles make a good representation of a local portion of a surface because three points with z values uniquely define a plane in three-dimensional space.

"Irregular" identifies the key advantage of TINs for surface modeling—points can be sampled with variable density to model areas where change in surface relief is abrupt.

"Network" reflects the topological structure that is implicit in a TIN. This structure enables sophisticated surface analysis as well as compact representation of a surface.

Creating TINs

TINs are made from mass points, which are points with elevations collected from a variety of sources. TINs are commonly compiled with photogrammetric instruments that sample elevations from pairs of aerial photographs precisely aligned in a stereo model. TINs are also produced from survey data, digitized contours, rasters with z-values, point sets in files or databases, or operations on other TINs.

From these input points, a triangulation is performed on the set of points. In a TIN, the triangles are called

faces, the points become *nodes* to a face, and the lines of faces are called *edges.*

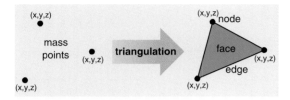

Each face in a TIN is a part of a plane in three-dimensional space. All of the faces in a TIN meet their neighbors precisely at each node and along each edge. Faces cannot intersect each other.

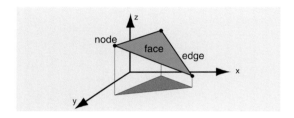

Triangulation and topology

Given a set of points, many possible triangulations can be created. ArcInfo uses an algorithm called the *Delaunay triangulation* to optimize how faces model a surface.

The basic idea of this algorithm is to create triangles that collectively are as close to equilateral shapes as possible. This keeps the interpolation of elevations at new points in closer proximity to the known input points.

A triangulation can be made from an input set of surface features represented by points, lines, and areas. First, a triangulation is made from points. Next, lines are inserted into the triangulation and new nodes are created wherever those lines split faces. Finally, areas are inserted; these can also split or clip faces.

After triangulation is complete, the TIN stores a list of nodes for each face, and for each face, a list of neighboring faces. This representation is similar to the topology represented by planar topologies. The difference is that nodes have elevations and faces must be triangles instead of arbitrary polygons.

Topology and triangulation

The Delaunay Triangulation

A Delaunay triangulation uses an algorithm to optimize the surface representation.

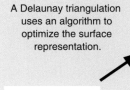

This triangulation fails the Delaunay test because the circle bounding nodes 1, 3, and 4 includes node 2.

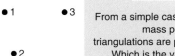

From a simple case of four mass points, two triangulations are possible. Which is the valid TIN?

The definition of the Delaunay triangulation specifies that any circle around three nodes in a triangle will not include any other node.

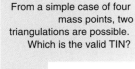

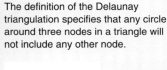

This triangulation satisfies the Delaunay test because a circle around each triangle contains no other nodes. This is the valid triangulation.

Topology in a TIN

A TIN is a topological data structure that manages information about the nodes that comprise each triangle and the neighbors to each triangle.

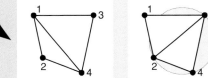

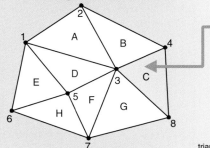

Triangle	Node list	Neighbors
A	1, 2, 3	–, B, D
B	2, 4, 3	–, C, A
C	4, 8, 3	–, G, B
D	1, 3, 5	A, F, E
E	1, 5, 6	D, H, –
F	3, 7, 5	G, H, D
G	3, 8, 7	C, –, F
H	5, 7, 6	F, –, E

Triangles always have three nodes and usually have three neighboring triangles. Triangles on the periphery of the TIN can have one or two neighbors.

You can create a TIN by entering surface features that represent elements of terrain such as point elevations, peaks, streams, and ridges.

Point elevations are the predominant input into a TIN and form the overall shape of the surface. They can be input from contour lines if necessary, but it is better to use points collected with photogrammetric devices because the operator can do a better job of visually sampling points that reflect terrain relief.

Streams, ridges, and similar surface features are then added to refine the surface model and sharpen the changes in relief. These features are preserved in the TIN and increase the model accuracy.

REPRESENTING SURFACE MORPHOLOGY

This is a view of surface features as shown on many maps. These features can be converted into a TIN.

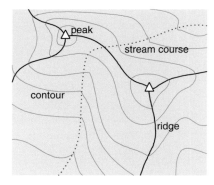

Point surface features

Mass points represent points at which a z-value is measured. After triangulation, they are preserved as nodes with the same location and elevation.

Line surface features

Breaklines are linear features that represent natural features such as courses and ridges or man-made features such as roadways. There are two kinds of breaklines: hard and soft.

Hard breaklines represent a slope discontinuity such as a stream course. While the surface is always continuous, its slope may not be. Hard breaklines preserve surface sharpness and improve the analysis and display of a TIN.

Soft breaklines let you add edges to represent line features, but do not represent a slope discontinuity. For example, you might want to add the imprint of a road to your surface model, but it may not significantly alter the local surface slope.

Areal surface features

Polygons represent objects such as lakes or coasts.

Replace polygons assign one constant z-value to the boundary and all interior heights.

Erase polygons mark all areas within the polygon as being outside the zone of interpolation for the mode. Analytic operations such as volume calculations, contouring, and interpolation will ignore these areas.

Clip polygons mark all areas outside the polygon as being outside the zone of interpolation for the model.

Fill polygons assign an integer attribute value to all faces within the polygon. No height replacement, erasing, or clipping takes place.

FUNCTIONAL SURFACES

A TIN is a surface representation with a single z-value for each point. An interesting quality of the TIN data representation is that it represents points in three-dimensional space, but the topological network of faces is constrained to two dimensions.

For this reason, the TIN data representation is sometimes said to model "two-and-a-half" dimensions. This description is not quite accurate, but illuminates the fact that surfaces have points measured in three dimensions but each point can have only one z-value.

Consequently, TINs are an example of a single-value function—given an input location, only one z-value can be interpolated. A minor constraint of TINs is that they cannot model the infrequent occurrences of negative slope, such as overhanging cliffs and caves. Unless you are modeling climbing routes in Yosemite Valley or passageways through Carlsbad Caverns, this is thankfully not a significant limitation.

Using surface features to create a TIN

Surface features

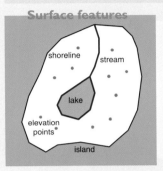

shoreline
stream
lake
elevation points
island

TINs are created from surface features that can be categorized by dimension: points, lines, and areas. These surface features are called mass points, breaklines, and polygons, and can be input from a variety of data sources.

→ triangulation

As surface features are input to a TIN, they become a topological network of nodes, edges, and faces. Each node in a TIN has a z-value. A hull is a boundary of surface interpolation. A TIN has at least one hull on its exterior; additional hulls can delimit islands and lakes.

TIN

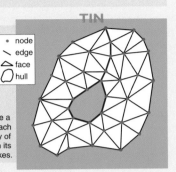

- • node
- \ edge
- △ face
- ◯ hull

When you build a TIN, mass points, breaklines, and polygons are progressively added to create and refine the surface model. At any time, you can add additional surface features to further improve the model.

Point surface features

Mass points represent locations with known z-values. They can represent specific features such as peaks or the bottom of a depression, but most are points sampled from a surface with measured z-values. The density of mass points should vary by the degree of change in the surface.

Mass points

From the set of mass points, the Delaunay triangulation algorithm is applied to create an initial TIN. This TIN reflects the overall shape of the surface, but does not yet well represent sharp changes in terrain such as streams and ridges.

Linear surface features

Breaklines represent natural features such as ridges, or man-made features such as roads. Breaklines can be identified as hard or soft, which influence how elevation isolines (contours) are drawn on a map.

Breakline

When breaklines are added, new nodes are created at the intersections with the initial edges, if those edges represent break edges. The TIN is updated to incorporate these new nodes.

Areal surface features

Polygons either represent areal features with a constant elevation, such as a lake or ocean, or are used to delimit the project boundary and clip or erase portions of the triangulation. Polygons can also be identified as hard or soft.

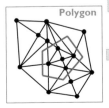

Polygon

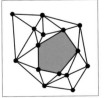

The TIN has now been refined to model areas of constant elevations and the boundary of interpolation.

After the TIN has reached this stage, you can inspect it and add any additional mass points, breaklines, or polygons to correct and improve the TIN surface model.

Surface features in a sample TIN

This illustration depicts a fictional TIN designed to show some of the key types of surface features as they are drawn in ArcMap.

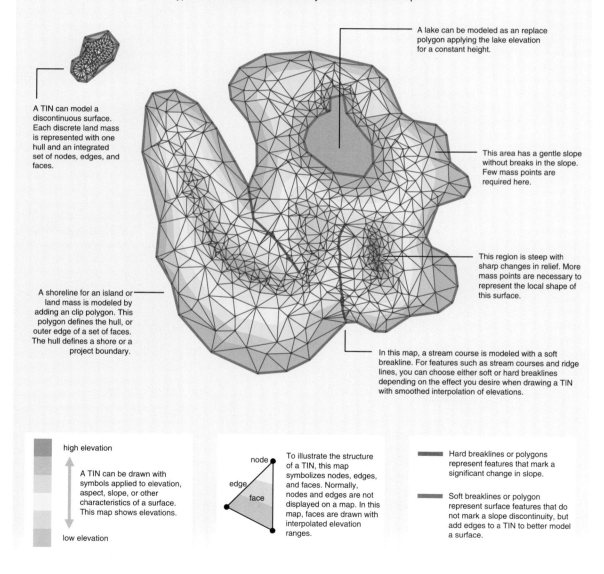

A lake can be modeled as an replace polygon applying the lake elevation for a constant height.

A TIN can model a discontinuous surface. Each discrete land mass is represented with one hull and an integrated set of nodes, edges, and faces.

This area has a gentle slope without breaks in the slope. Few mass points are required here.

This region is steep with sharp changes in relief. More mass points are necessary to represent the local shape of this surface.

A shoreline for an island or land mass is modeled by adding an clip polygon. This polygon defines the hull, or outer edge of a set of faces. The hull defines a shore or a project boundary.

In this map, a stream course is modeled with a soft breakline. For features such as stream courses and ridge lines, you can choose either soft or hard breaklines depending on the effect you desire when drawing a TIN with smoothed interpolation of elevations.

high elevation

A TIN can be drawn with symbols applied to elevation, aspect, slope, or other characteristics of a surface. This map shows elevations.

low elevation

node

edge

face

To illustrate the structure of a TIN, this map symbolizes nodes, edges, and faces. Normally, nodes and edges are not displayed on a map. In this map, faces are drawn with interpolated elevation ranges.

Hard breaklines or polygons represent features that mark a significant change in slope.

Soft breaklines or polygon represent surface features that do not mark a slope discontinuity, but add edges to a TIN to better model a surface.

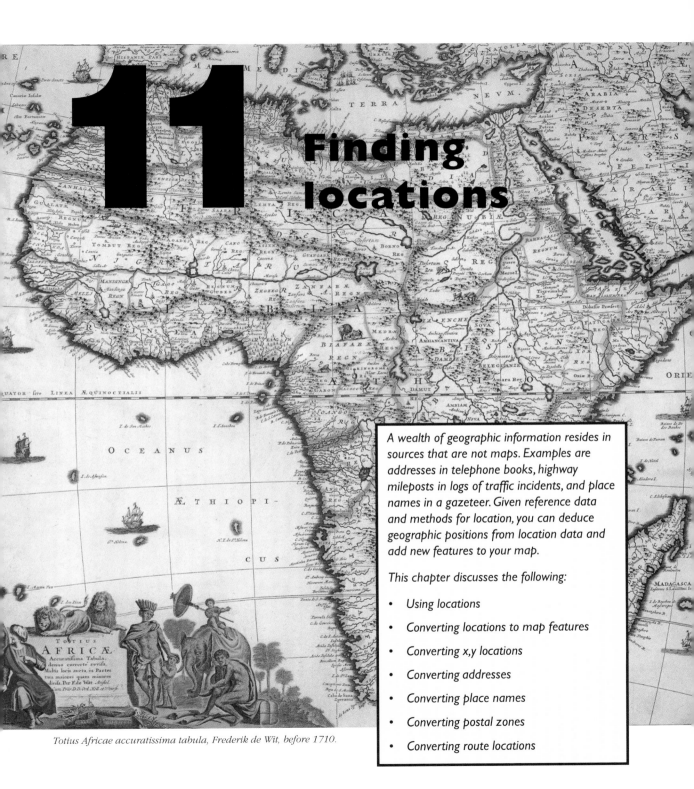

11 Finding locations

A wealth of geographic information resides in sources that are not maps. Examples are addresses in telephone books, highway mileposts in logs of traffic incidents, and place names in a gazeteer. Given reference data and methods for location, you can deduce geographic positions from location data and add new features to your map.

This chapter discusses the following:

- Using locations
- Converting locations to map features
- Converting x,y locations
- Converting addresses
- Converting place names
- Converting postal zones
- Converting route locations

Totius Africae accuratissima tabula, Frederik de Wit, before 1710.

The majority of information that people track includes some sort of reference to a geographic location. For example:

- A store keeps records of customers. This information includes their addresses and the store location from which they have made purchases.

- A police department keep logs of crime incidents. These records reference locations of transgressions by street intersections or addresses.

- A department of transportation keeps track of road maintenance through a mileposting system.

A unique quality of a geographic information system is its ability to integrate diverse types of data into a common geographic framework. Tying diverse data together gives you considerable freedom to explore the relationships between entities such as people, highways, land, stores, and natural features.

This chapter discusses the types of locations, how to define location methods and set up reference data to infer geography, and the addition of point, line, or area features to your map from location data.

TAXONOMY OF LOCATION DATA

Geographic locations come in many forms. The most common locations are street addresses, but other locations include mileposts, names of significant places or features, and various types of grid locations.

These are the types of geographic locations from which you can derive features and add to your geodatabase.

x,y coordinates

Often, an organization will maintain tables that contain records with attributes for x,y coordinates.

For example, an environmental agency might create records of air quality readings. For each reading, a coordinate can be collected with a GPS receiver or read from a map. To study these readings in the context of sources of air pollution, you can convert the collected coordinates and create points in a new feature class. You can compare these point features with locations of factories to analyze the cause and dispersion of pollutants in the environment.

Street addresses

Many databases contain addresses to keep track of customers and business locations. For many countries, accurate street maps can be combined with methods to convert an address to a location.

For example, a corporation can maintain databases of customer purchases that contain the customer address. The corporation can convert the addresses to points on a map to study the optimum geographic placement of new stores.

Postal codes

Sometimes, databases might not contain addresses but postal codes, such as the ZIP Code system in the United States. These codes can reference one to several thousand addresses.

For example, consumer demographic data is widely available aggregated on postal zones. Someone who wants to focus a marketing campaign on a desired demographic can perform spatial clustering analysis to optimize the cost benefit of advertising.

Place names

People find their way by making reference to landmarks, such as government buildings, schools, and mountain peaks.

For example, a tourist bureau might develop informational kiosks that let the user select from a database of hotels, restaurants, and points of interest. A database with place names and positions can be used to construct maps to guide the user.

Route locations

Many of the things that are built—roads, canals, railroads—are designed with a linear measurement system based on distances from starting points.

For example, a department of transportation keeps extensive records of all aspects of a highway system, such as pavement quality, incidents of accidents, and signage. Data for these entities is stored in a mileposting system that references a route and linear distance from a known point. Given route reference data in the form of accurate street maps, any route measurement can be added to a map.

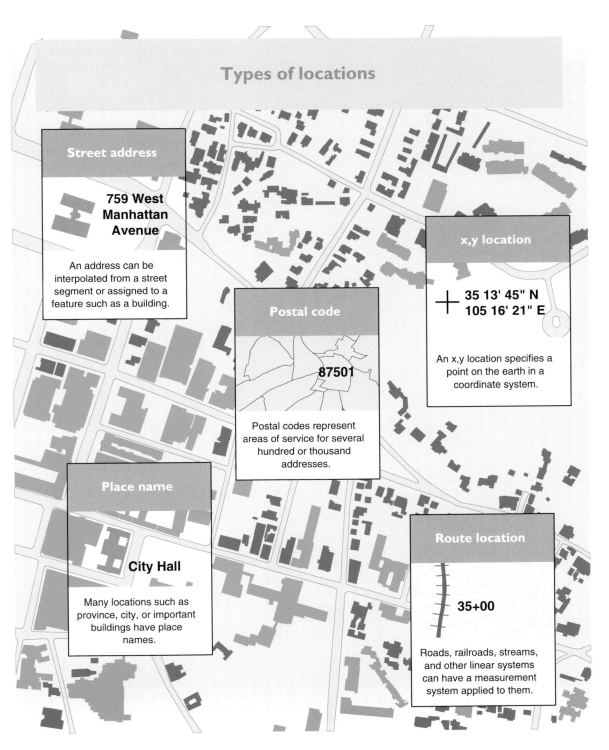

Types of locations

Street address

759 West Manhattan Avenue

An address can be interpolated from a street segment or assigned to a feature such as a building.

x,y location

$+$ **35 13' 45" N 105 16' 21" E**

An x,y location specifies a point on the earth in a coordinate system.

Postal code

87501

Postal codes represent areas of service for several hundred or thousand addresses.

Place name

City Hall

Many locations such as province, city, or important buildings have place names.

Route location

35+00

Roads, railroads, streams, and other linear systems can have a measurement system applied to them.

Using *locators*, you can convert a location to a known position on a map.

If you can take some kind of location and figure out a precise or approximate position on your own, you can train ArcInfo to do the same.

The general technique for adding locations to a map is to prepare a locator, apply location data to it, and use the point, line, or polygon features it creates on a map.

LOCATORS

A locator is a combination of reference data and location method. There are locators for each type of location.

For addresses, the reference data is a street map that contains street centerline segments with address ranges, street names, and related attributes. The location method converts addresses to a position along the street segment.

For x,y locations, the reference data is the definition of the coordinate system. The location method is the coordinate transformation function built into ArcInfo.

For place names and postal codes, the reference data is a map with place names or codes as an attribute. The location method is a simple match between the input place name or code and the feature that contains it.

For route locations, the reference data is a street map with routes and a measurement system. The location method finds the input route, finds the line that

contains the input measure, and calculates the position of the input measure.

Location data

The location data is a table with one or more prescribed fields that describe the locations you want to find on a map. Location data is applied to a locator to create points on a map.

Location feature classes

The result of the application of a locator is a new feature class that contains points, lines, or polygons.

Depending on your application, you can create and store the feature class or you can create temporary features each time your application reads the locations. The location feature class has a spatial reference that is defined by the locator.

Custom locators

ArcInfo presents a number of standard locators that are ready to use. However, some applications demand sophisticated custom location methods. An example would be an address locator for a country that does not follow the postal convention of house numbers ascending on a range along a street segment. For this circumstance, you might find custom locators developed by international distributors or GIS consulting firms.

The location framework in ArcInfo is open, flexible, and extensible. You can define as many custom locators as you want.

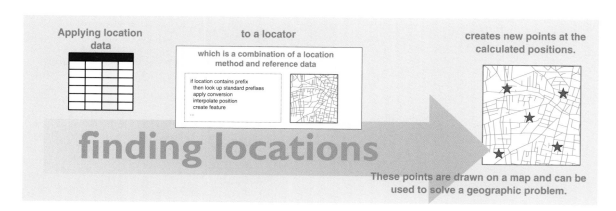

Applying location data to a locator which is a combination of a location method and reference data creates new points at the calculated positions.

if location contains prefix
then look up standard prefixes
apply conversion
interpolate position
create feature
...

finding locations

These points are drawn on a map and can be used to solve a geographic problem.

You can convert a database table containing x,y values into points on a map. These points are created in a new feature class that contains all of the attributes of the input database table.

Some common scenarios for x,y locations are points collected when sampling environmental data, performing field maintenance of structures, surveying corners of a land parcel, and tracking the motion of a boat, airplane, or other vehicle.

The increasing use of GPS receivers makes x,y locations more common as attributes in database tables.

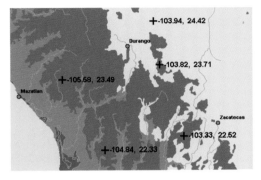

Points collected from a field survey drawn on a map

Input location table

The x,y values in the input location table must be in two numeric fields. The x,y values cannot be combined into one field. The fields with x,y values can have any name.

The implicit coordinate system of the x,y values must be in a form that the ArcInfo spatial reference system can accept.

Locations and spatial reference

It is not necessary that the x,y values be in a spatial reference already present in your geodatabase. For example, x,y data is commonly collected as latitude/longitude pairs. When you create a locator for x,y locations, you will specify the coordinate system of the input data and a spatial reference for the newly created feature class.

Therefore, records with latitude/longitude values can be converted into a feature class with a spatial reference that can be directly overlaid on other feature datasets and feature classes in your geodatabase.

The x,y locator will check the input x,y values to ensure that they fall within the expected range of the spatial reference system. For example, longitude values must be between −180 and 180.

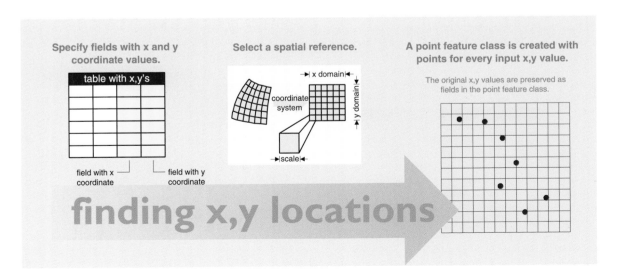

The single most common form of geographic information is the address. There are many more addresses kept in corporate and government database tables than features in all the digital maps ever created. All businesses and governments keep track of people directly or indirectly through their addresses.

Because addresses are so prevalent in any information system, an important area of work for GIS professionals is developing and updating national street maps with address attributes and methods for finding addresses.

ArcInfo comes with a CD–ROM containing address reference data for the entire United States. ArcMap includes a set of predefined address locators designed to work with this data. Every address in the United States can be matched against this data—this makes the job of finding positions for addresses in your tables easier. Your main job will be identifying and correcting errors in your address tables.

If you are outside the United States, check with your international ESRI distributor for the availability of national street data with suitable address locators.

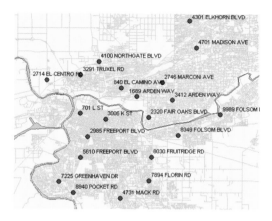

A set of addresses converted to annotated features on a map

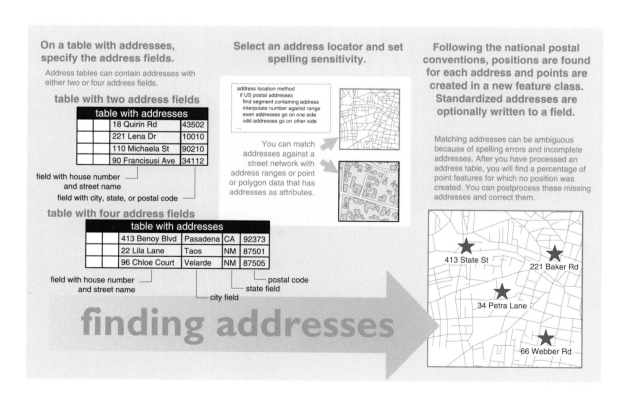

On a table with addresses, specify the address fields.

Address tables can contain addresses with either two or four address fields.

table with two address fields

		table with addresses	
		18 Quirin Rd	43502
		221 Lena Dr	10010
		110 Michaela St	90210
		90 Francisusi Ave	34112

field with house number and street name
field with city, state, or postal code

table with four address fields

		table with addresses			
		413 Benoy Blvd	Pasadena	CA	92373
		22 Lila Lane	Taos	NM	87501
		96 Chloe Court	Velarde	NM	87505

field with house number and street name
city field
state field
postal code

finding addresses

Select an address locator and set spelling sensitivity.

address location method
if US postal addresses
find segment containing address
interpolate number against range
even addresses go on one side
odd addresses go on other side
...

You can match addresses against a street network with address ranges or point or polygon data that has addresses as attributes.

Following the national postal conventions, positions are found for each address and points are created in a new feature class. Standardized addresses are optionally written to a field.

Matching addresses can be ambiguous because of spelling errors and incomplete addresses. After you have processed an address table, you will find a percentage of point features for which no position was created. You can postprocess these missing addresses and correct them.

MATCHING ADDRESSES TO STREETS

The reference data for matching addresses to streets consists of a set of lines representing street segments between intersections. Each line has a number of attributes such as street name and address ranges.

These are the components of addresses you might find in street reference data:

- Address ranges represent progressive numbering of houses along a street segment. Most locales have address ranges on the left and right sides.

- A prefix direction denotes a direction in reference to a local center of addressing. For example, a city may have these two addresses: 20 West Main Street and 20 East Main Street. The prefix direction, "West" or "East," might be necessary to unambiguously locate an address.

- A street name is the main identifier for a street segment. Examples are "Main" and "Wilshire."

- Each street has a type. Examples are "Road" and "Street." Sometimes, street types are prefixes to street names, such as "Highway" and "Avenida."

- Addresses for some cities have suffix directions, such as "NW" or "SE."

- Street reference data can contain the left and right postal zones for each street segment. This information is used to validate address matching.

Street data with left and right address ranges

Street reference data can contain right and left address ranges. The reference data on the U.S. street data CD–ROM is organized in this way.

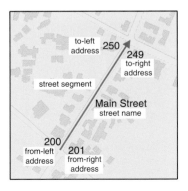

The United States and some other countries follow a postal convention that odd addresses are located on one side of a street and even addresses on the other.

An advantage of using left and right address ranges is that positions found for addresses can be placed correctly on the left or right side of a street.

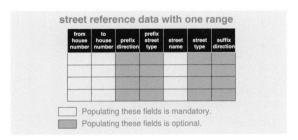

These are the fields required by the standard address locator for data with right and left ranges.

Street data with single address ranges

For some locales, address ranges are available for the beginning and end of each street segment. When positions are found for addresses, they are placed on the street segment.

These are the fields required by the standard address locator for data with a single address range.

Processing addresses

When the address locator processes an address table, a match score is calculated for each point. If the match score is greater than the threshold you have defined, a point is placed on the map. If not, a point is still created with a null position that you can review to correct addresses. Some causes of low match scores are incomplete addresses, misspelled street names, or house numbers out of range.

Most of the time, you will use conventional addresses in your address tables; however, two additional techniques match addresses based on place names and street intersections.

FINDING PLACES BY LINKING ADDRESSES

An address table can contain place names. If you have a place name table, you can direct the address locator to first locate a matching place name in the place name table, and then a matching address in the address reference data.

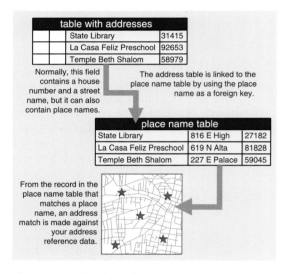

table with addresses			
		State Library	31415
		La Casa Feliz Preschool	92653
		Temple Beth Shalom	58979

Normally, this field contains a house number and a street name, but it can also contain place names.

The address table is linked to the place name table by using the place name as a foreign key.

place name table			
State Library	816 E High	27182
La Casa Feliz Preschool	619 N Alta	81828
Temple Beth Shalom	227 E Palace	59045

From the record in the place name table that matches a place name, an address match is made against your address reference data.

This is a two-level matching process.

FINDING STREET INTERSECTIONS

The intersection of two streets marks a point on a map. You can add street intersections to an address table and find the position of the crossing.

Using the same address reference data as they would for address matching, the standard address locators can take combinations of streets and locate their intersections on a map.

Street intersections are specified by the use of a connector such as "AND", "&", "/", or other. You can define a set of valid connectors for your address tables, but you should take care that they are not otherwise used as a component of a street name.

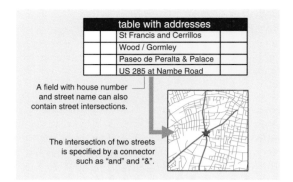

table with addresses	
	St Francis and Cerrillos
	Wood / Gormley
	Paseo de Peralta & Palace
	US 285 at Nambe Road

A field with house number and street name can also contain street intersections.

The intersection of two streets is specified by a connector such as "and" and "&".

This technique creates new points for intersections.

FINDING BUILDING ADDRESSES

Another technique is to match addresses to reference data that contains points or polygons representing buildings, and that has address fields.

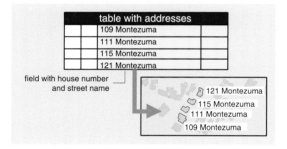

table with addresses	
	109 Montezuma
	111 Montezuma
	115 Montezuma
	121 Montezuma

field with house number and street name

121 Montezuma
115 Montezuma
111 Montezuma
109 Montezuma

This address reference data is a point or polygon feature class that contains several address fields. Just as with other addresses, you will specify a match score to set a threshold for qualifying matched addresses.

building reference data

The standard address locator uses these fields in the building reference data.

house number	prefix direction	prefix street type	street name	street type	suffix direction	zone

☐ Populating these fields is mandatory.
▨ Populating these fields is optional.

Points are created for matches on a point reference feature class. Polygons are created for matches on a polygon reference feature class.

On the previous page, you saw how place names can be located through a two-level matching process—first to a place name table and then to address reference data based on street segments with address ranges.

Instead of a street network with address ranges, you might instead have reference data with points or polygons and place names as attributes. For example, you might have a nationwide map of all the counties or local jurisdictions. This reference data would not naturally contain addresses; those entities are too large for an address to be meaningful.

A place name can have one part or two parts. For example, you could divide county names into two parts: the name and the county type. Examples of county types in the United States are "County," "Parish," and "Borough."

If the place name table has one-part names, the place name reference data must have one field for place names. Similarly, if the place name table has two fields for place names, the place name reference table must also have two fields.

If you match a place name table against a point feature class, the new feature class will contain points. If you match a place name table against a polygon feature class, the new feature class will contain polygons.

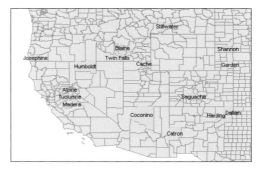

A set of place names matched to a set of polygons representing counties

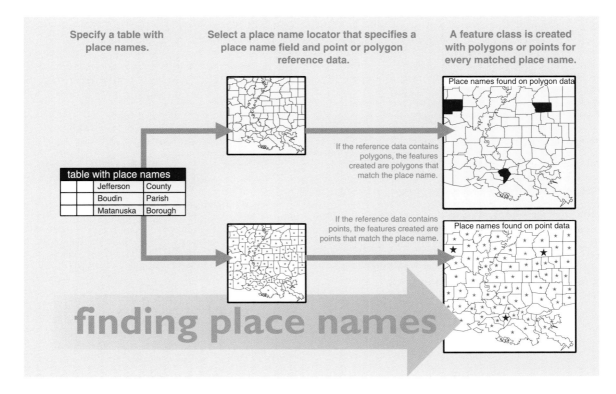

Specify a table with place names.

Select a place name locator that specifies a place name field and point or polygon reference data.

A feature class is created with polygons or points for every matched place name.

table with place names	
Jefferson	County
Boudin	Parish
Matanuska	Borough

Place names found on polygon data

If the reference data contains polygons, the features created are polygons that match the place name.

If the reference data contains points, the features created are points that match the place name.

Place names found on point data

finding place names

If you are matching postal codes in the United States, you can use a ZIP Code locator.

A ZIP Code has two parts: a five-digit prefix that identifies an area served by a post office that may have several thousand addresses, and a four-digit extension that identifies individual addresses.

You can match either type of ZIP Code. For ZIP+4 codes, you can choose from three formats: as one number, such as "875051357"; with a space, such as "87505 1357"; or with a hyphen, such as "87505-1357."

Using ZIP+4 codes will give you the most accurate position because you will be matching to reference data with a greater number of points. Using ZIP+4 ranges will give you less accuracy because you will be matching to reference data with fewer points. Matching five-digit ZIP Codes will be the least accurate because you will match to the centroid of the five-digit ZIP area.

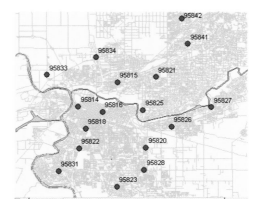

A set of ZIP Codes matched to the centroids of polygons representing ZIP Code areas.

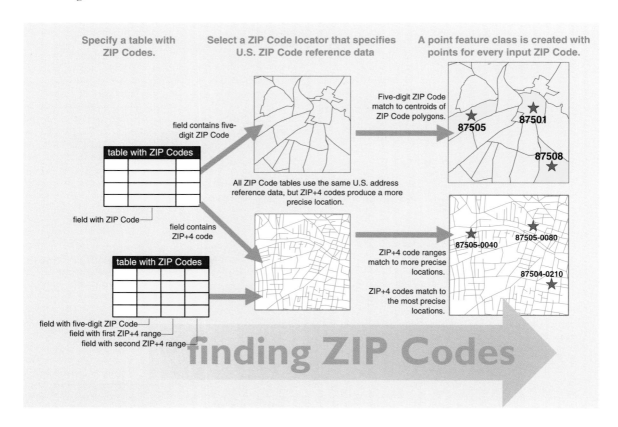

Specify a table with ZIP Codes.

Select a ZIP Code locator that specifies U.S. ZIP Code reference data

A point feature class is created with points for every input ZIP Code.

field contains five-digit ZIP Code

table with ZIP Codes

field with ZIP Code

All ZIP Code tables use the same U.S. address reference data, but ZIP+4 codes produce a more precise location.

Five-digit ZIP Code match to centroids of ZIP Code polygons.

87505   87501

87508

field contains ZIP+4 code

table with ZIP Codes

field with five-digit ZIP Code
field with first ZIP+4 range
field with second ZIP+4 range

ZIP+4 code ranges match to more precise locations.

ZIP+4 codes match to the most precise locations.

87505-0040   87505-0080

87504-0210

finding ZIP Codes

Certain locations are measures along linear systems. For example, road maintenance data kept by a department of transportation references measurements of locations along routes.

A route is a collection of polylines that have a common identifier and contain measures. (To review polylines and measures, read chapter 6, "The shape of features.")

A route location can represent either a point along a route or a segment along a route between two points.

A route location representing a point on a route contains a route ID and a single measure value. The route ID specifies the polylines with measures to search for. The measure value identifies the point that is interpolated on measures of one polyline.

A route location representing a segment along a route has a route ID and two measure values. The segment is interpolated between the two measure values, and a line feature class contains the resulting polylines.

The route reference data is a line feature class with route IDs assigned to polylines and measures established to each point in the polyline.

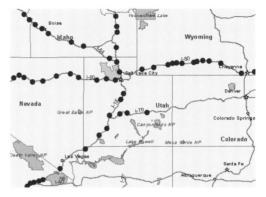

Points representing traffic accident locations.

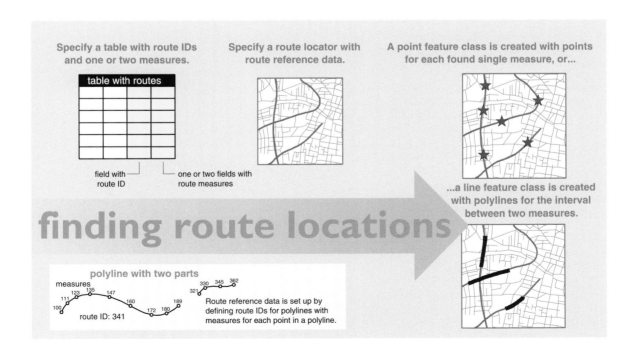

Specify a table with route IDs and one or two measures.

table with routes

field with route ID — one or two fields with route measures

Specify a route locator with route reference data.

A point feature class is created with points for each found single measure, or...

...a line feature class is created with polylines for the interval between two measures.

finding route locations

polyline with two parts

measures
123 135 147
111 160 189
100 172 180
route ID: 341
330 345 362
321

Route reference data is set up by defining route IDs for polylines with measures for each point in a polyline.

12

Geodatabase design guide

The geographic data model, implemented in a geodatabase design, is the foundation for all activities with a GIS—creating expressive maps, retrieving information, and performing spatial analysis. Designing a geodatabase to meet these goals is a deliberate process.

These are the topics in this chapter:

- Purpose and goals of design

- Overview of design steps

- Step 1: Model the user's view

- Step 2: Define entities and relationships

- Step 3: Identify representation of entities

- Step 4: Match to ArcInfo data model

- Step 5: Organize into geographic datasets

Generalis totius Imperii Russorum novissima tabula, Johann Baptist Homann, before 1724.

A geographic information system has the potential to help your organization accomplish a myriad of tasks, from daily operations to long-term planning. Effectively implementing your GIS allows you to realize this potential, offering efficient ways to perform functions, store and share data between organizational units, and integrate with other technologies.

This chapter reviews the general steps to achieve your own geodatabase design.

THE NEED FOR DESIGN

What makes GIS implementation effective is a good database design. And what makes a database design good is asking the right questions:

- How can GIS technology be implemented to streamline existing functions, or change the way a particular goal is achieved?

- What data will benefit the organization most?

- What data can be stored?

- Who is, or should be, responsible for maintaining the database?

How you answer these questions will deepen your understanding of GIS technology, as well as provide new insight into your organization and its functions.

Design for GIS implementation is like any other design. It starts with understanding goals and progresses through increasing levels of detail as information is gathered and you approach implementation.

An example of this is a transportation model that begins by studying existing traffic flow and applying patterns of population growth. This model would be developed in a series of steps with progressively more detail. The preliminary proposal and budget evolves to detailed engineering drawings.

Because it is time consuming and produces no end-use applications, the design process often receives little attention, if any. There are risks associated with avoiding design. If you do not go through the design process, you risk having a poorly constructed database that does not meet your requirements, now or in the future. This can result from duplicate, missing, or unnecessary data; inappropriate representation of data; or lack of proper data management techniques.

This section focuses on database design; however, you will quickly realize that the database and the applications it serves cannot be treated entirely independently. As you progress through the database design, you should also define the applications that will create, use, and manage the data.

OBJECTIVES OF DESIGN

Design is the process in which goals are defined, design alternatives are identified, analyzed, and evaluated, and an implementation plan is agreed upon. At the highest level, the design provides a picture of where you are, where you are going, and how to get from one place to the other. As you progress through the design, you increase detail, adding data definitions and assigning the appropriate ArcInfo spatial data structures.

A database design provides a comprehensive architecture for the database. The design allows you to view the database in its entirety and evaluate how the various aspects of it need to interact. Expending time and money to identify and resolve design issues early saves having to expend greater resources later trying to solve what may well have become insurmountable problems.

A good design results in a well-constructed, functionally and operationally efficient database that:

- Satisfies objectives and supports organizational requirements.

- Contains all necessary data but no redundant data (unless explicitly planned and properly documented).

- Organizes data so that different users access the same data.

- Accommodates different views of the data.

- Distinguishes applications that maintain data from those that use it.

- Appropriately represents, codes, and organizes geographic features.

Design is time consuming and intensive, but you will benefit from:

- Increased flexibility of data retrieval and analysis.

- Increased likelihood of users developing applications.

- Distributed cost of data capture, storage, and use.

- Facilitated data that supports many different uses.

- Maintained data that supports many different users.

- Extensibility that readily accommodates future functionality.

- Minimized data redundancy.

DESIGN GUIDELINES

The design process can be quite substantial. Here is some advice to ease the process and help ensure success:

- Involve users. By contributing, they will gain a sense of ownership and you will gain invaluable knowledge for your geodatabase design.

- Take it one step at a time. It is not necessary to create a complete detailed design all at once; design is an interactive and iterative process. You can progress in stages as appropriate for the needs of your organization.

- Build a team. A wide range of information, skills, and decision making is required during this process. At different stages, your team will comprise various experts throughout your enterprise.

- Be creative. The initiation of a new project is a good opportunity to survey new technology and processes. There is considerable potential to enhance how GIS serves your organization's goals and objectives.

- Create deliverables. It is best to divide a large project into discrete and identifiable units of work. Project milestones should be defined to be

no less frequent than two months or so. This will keep your project focused and earn management support.

- Keep organizational goals and objectives in focus. It is essential that the design and implementation process always be focused on the real requirements of your organization and its customers.

- Do not add detail prematurely. Add detail at the appropriate step. For example, do not try to define all of the validation rules for feature classes before geodatabases are constructed. Selectively introduce implementation details throughout the project so that the team can progress to the next step.

- Document carefully. The more complex the environment, the greater the benefit from documenting your design. The use of business-diagramming software is especially useful to communicate your design.

- Be flexible. The initial design will not be the final design as implemented. The design will evolve as your organization changes, new technology is introduced, and people become more adept with the technology.

- Plan from your model. Create an implementation plan that addresses your organization's key priorities in a manageable fashion. If you need to create new datasets, build the data management applications first.

This database design process is presented as a series of steps. While by-products of your design activities can include identifying applications, identifying educational and training requirements, and setting standards for data collection and maintenance, only database design is covered in this section.

The process is not meant to present a formal methodology that is beyond the scope of this section. The intent is more to guide you through a design if you do not already use a formal methodology.

The steps are:

- Model the user's view.

- Define entities and their relationships.

- Identify representation of entities.

- Match to the geodatabase data model.

- Organize into geographic datasets.

The first three steps develop the conceptual model, classifying features based on an understanding of the data required to support the organization's functions, and deciding their spatial representation (point, line, area, image, surface, or nongeographic). The last two steps develop the logical data model, matching the conceptual models to ArcInfo geographic datasets.

Steps in building a geodatabase

Model the user's view of data.

Identify organizational functions.
Determine data needed to support functions.
Organize data into logical groupings.

Define objects and relationships.

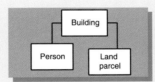

Identify and describe objects.
Specify relationships between objects.
Document model in diagram.

Select geographic representation.

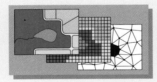

Represent features with points, lines, and areas.
Characterize continuous phenomena with rasters.
Model surfaces with TINs or rasters.

Match to geodatabase elements.

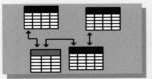

Determine geometry type of discrete features.
Specify relationships between features.
Implement attribute types for objects.

Organize geodatabase structure.

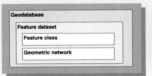

Organize systems of features.
Define topological associations.
Assign coordinate systems.
Define relationships and rules.

The objective of this step is to ensure a common understanding between the design team and those who have a vested interest in the implementation of your GIS.

In this step, you:

- Identify the functions that support the organization's goals and objectives.
- Identify the data required to support the functions.
- Organize the data into logical sets of features.
- Define an initial implementation plan.
- Identify organizational functions.

An anticipated benefit of your GIS implementation is improvement in the way your organization conducts its business.

IDENTIFY ORGANIZATIONAL FUNCTIONS

An organization performs business functions that address its goals and objectives. These functions are the starting point for your database design. You work with business functions rather than organizational units, because functions are more stable than organization. That is, a function performed by one department today may be performed by another department next year.

To begin, identify each of the functions within scope of your project. For each function identified, provide a general description of the activities that fall within that function. Activities may include managing the land development approval process, controlling land use, and developing agreements for infrastructure construction by a developer.

Apart from the users themselves, documents and maps serve as good information sources. Look for general publications, strategic plans, and information systems plans.

LOCATE DATA SOURCES

Once the functions are compiled, identify the data that supports them. Determine whether the function "creates" or simply "uses" the data.

In general, you work with two kinds of data: the data of interest in your field and background data.

Naturally, the data of interest will be modeled in more detail.

You can analyze each function's scope by examining interactions with other functions and external players. Most often, data that flows out from the function has been created by the function. This indicates that it is responsible for the definition, collection, storage, and distribution of that data.

Data that flows into the function is generally the responsibility of another function, though data received from an external organization may be stored and managed internally. Exchanges are in many forms, including data, guidelines, requests, and responses.

The question to answer at this stage is, "Who or what does this function interact with and what is the nature of that interaction?"

In relating data to the functions that create and store them, you may discover synonyms, polynyms, and functions that duplicate the capture and storage of data. These situations should be resolved immediately or at least kept in a log for future resolution.

This should be an interactive step with those who perform the function—after all, they know what they do, who they interact with, and what information is exchanged. After documenting the required data, be sure to give them an opportunity to validate the diagram and any supporting text.

ORGANIZE DATA INTO LOGICAL GROUPINGS

Make a top-level grouping of all the data you expect to interact with in your GIS. These groupings represent systems such as "water utility," "land records," "streets," and "terrain."

Each of these groupings is operated by a function to either receive or transmit information. An example is that a surface model with rainfall amounts transmits hydrological data to a stream network.

Each of these groupings should have a common coordinate system, topological type (network, planar, or none), and generally interact with each other.

Model the user's view of data

Identify organizational functions

The geodatabase design will be influenced by the structure of your organization. Distinct departments may have responsibility for different segments of the geographic data.

At a basic level, you begin by identifying the providers and consumers of geographic information. The key data flows are modeled. This is the starting point for identifying logical groupings of data.

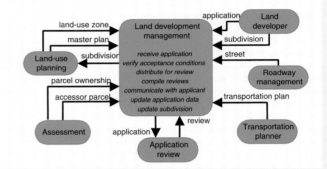

Determine data needed to support functions

For each function, identify all of the types of data that are necessary to fulfill this group's requirement to deliver information.

Land records	
Types of data	Data source
Parcel	Subdivision plats
Easement	Engineering records
Parcel description	Land title
Parcel photograph	Historic archive
Owner	Land assessment
Address	Phone database

For each data type, identify the likely source of data. A part of the project plan must include an estimate for cost of data capture, processing, and validation.

Organize data into logical groupings

From an inventory of all the types of geographic data that an organization maintains, identify a modest set of groupings that comprise all of your geographic data systems.

The previous step determined the broad classification of functions, data, and the relationships between them. In this step, you examine the data classification more closely, identifying distinguishable objects, called entities, that have a common set of properties.

You will:

- Identify and describe entities.

- Identify and describe the relationships among these entities.

- Document the entities and relationships with UML diagrams.

It is recommended that you document this design using business graphics software such as Visio®. On this diagram, you would have boxes for entities and lines for relationships.

This step is significant because it adds detail to the user's view of the data they work with. It is most important that users be involved in the definition and validation of the models produced in this step.

You will deal with a lot of data during this step. To partition the task into manageable units, focus on one function at a time. This will guide which data to focus on. It may take several iterations to clarify the definitions of entities and their relationships.

Articulating entities and relationships

Identify entities and relationships by interpreting statements. Nouns tend to be entities while verbs define relationships between entities.

- *A valve controls the flow of gas.* This statement describes an entity.

- *A gas device connects to one or more gas lines.* This statement describes a structural relationship between entities.

- *A gas system is composed of gas devices and gas lines.* This statement describes the aggregation of entities to make a new, more complex entity.

- *A gas main is a type of gas line.* This statement describes a subclassification of entities.

Be aware of verbs masquerading as nouns (connection, description, identification, and aggregation). These tend to obscure the relationships.

Documenting entities and relationships

A concise and clear way to document this stage of the design is to create simple UML diagrams. Review the end of chapter 1, "Object modeling and geodatabases," for a quick primer on UML notation.

UML is used throughout this book and other diagrams available with ArcInfo to document the ArcInfo system architecture. UML is also appropriate for documenting your data model.

This is what a portion of your diagram might look like at this stage:

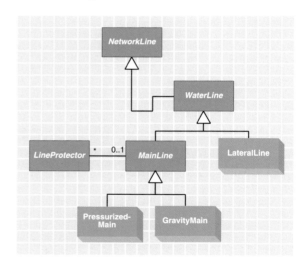

This diagram states the following:

- A water line is a type of network line.

- A main line and a lateral line make up a type of water line.

- A main line can be associated with zero to many line protectors. A line protector can be associated with zero or one main line.

- A pressurized main and gravity main are types of main lines.

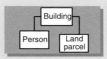

2

Define objects and relationships

identify entities and their relationships

entity	related to
Water utility	
Pump	–
Meter	–
Meter box	Meter
Valve	–
Water main	–
Treatment plant	–
Land records	
Parcel	–
Easement	–
Parcel description	Parcel
Parcel photograph	–
Owner	Parcel
Address	–
Streets	
Street	–
Bridge	–
Name	Street
Traffic light	–
Bus route	–
Bus stop	–
Environment	
Historic monument	–
Fence	–
Vegetation cover	–
Place names	–
River valley	–
Satellite image	–

Identify and describe objects

Form sentences that state the entities and their behavior. The nouns are entities and the verbs are relationships.

This step can be done by writing a progressive series of statements starting with "a water system is composed of devices and water lines." Each statement should be simple and accurate.

A valve controls the flow of water.

A water device connects to one or more water mains.

A water system is composed of devices and water lines.

A water main is a type of water line.

Specify relationships between objects

Many entities have close relationships with other entities. Relationships guide your geodatabase design.

A meter box is composed of meters.

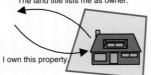

The land title lists me as owner.

I own this property.

A street name has a relationship with a street feature.

Document model in diagram

Once you have collected your list of entities and relationships, it is a good practice to create a data model diagram.

Using business graphics software, start by making boxes for entities and lines with arrows for relationships. This diagram will facilitate discussion with domain experts and advance the refinement of the model.

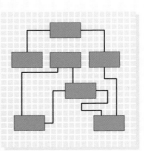

In this step, you classify entities by the type of representation. Some entities will have a geometric representation with corresponding attributes; these are classified by their geometric characteristics. Other entities will be represented by alphanumeric information only, still others by images, photographs, or drawings.

Consider whether:

- The feature might be represented on a map.

- The shape of the feature might be significant in performing geographic analysis.

- The feature is data that can be accessed and visualized through its relationship with another feature (for example, ownership information about a parcel can be accessed by selecting a parcel).

- The feature will have different representations at different map scales.

- Textual attributes of the feature will be displayed on the screen or map products.

The following terms are provided for assigning a type. The information developed during this step should be cataloged as part of the feature's data dictionary entry.

- Point—illustrates the location of a feature whose shape is too small to be defined as an area on a map of a given scale.

- Line—illustrates the location of a feature whose shape is too narrow to be defined as an area on a map of a given scale.

- Area—illustrates the location and polygonal shape of a feature on a map of a given scale.

- Surface—illustrates the shape of a feature as in an "area," but also includes shape resulting from changes in elevation.

- Raster—represents an area using rectangular cells (satellite image, aerial photograph, continuous data) and can be used for analysis.

- Image, photo, drawing—each represents a digital picture and cannot be used for analysis.

- Object—identifies a feature for which no point, line, or area is required, and for which there is no geometric or graphic representation.

If features could be represented in two forms depending on scale, identify both possibilities in the data dictionary, and use the more complex representation for consideration in the remainder of the analysis.

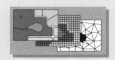 **3** Select geographic representation

set spatial representation
as vector, raster, and TIN

entity	related to	spatial type
Water utility		
Pump	–	point
Meter	–	point
Meter box	Meter	point
Valve	–	point
Water main	–	line
Treatment plant	–	point
Land records		
Parcel	–	area
Easement	–	line
Parcel description	Parcel	text
Parcel photograph	–	image
Owner	Parcel	object
Address	–	location
Streets		
Street	–	line
Bridge	–	point
Name	Street	text
Traffic light	–	point
Bus route	–	line
Bus stop	–	point
Environment		
Historic monument	–	point
Fence	–	line
Vegetation cover	–	area
Place names	–	text
River valley	–	surface
Satellite image	–	image

Represent discrete features with points, lines, areas

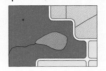

You can model the richest expression of features with the vector types. These entities are well defined on a map and are permanent.

point	*an entity too small to map with a line or area*
line	*a long entity too narrow to map with an area*
area	*an entity with length and width at the map scale*
annotation	*a descriptive label on an entity*
object	*a nongeographic entity, such as an owner*

Characterize continuous phenomena with images

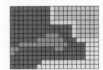

Images have versatile application in a GIS. You would specify images for aerial or satellite photographs, photographs of facilities, and any scanned documents.

image *a file that contains a continuous valued map, aerial photograph, copy of a plat, or picture of a building*

Model terrain with surfaces

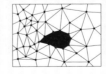

When you model a continuous phenomenon that has a z value, specify surface. (Later, you will decide whether TIN or raster is better for the surface.)

surface *a system of points or locations with elevation values that form a mesh for a mathetical approximation of the shape of the earth*

The objective of this step is to determine how data is to be represented in ArcInfo. For each of the spatial types identified in the previous step, you now assign a corresponding ArcInfo representation.

The focus now turns from understanding the user requirements to developing an efficient and effective database schema. It is important that the team have members who understand the geodatabase data model and analysis capabilities as well as other data management technologies to be used for your database.

In this step, you:

- Determine the appropriate geodatabase representation for entities.

- Ensure that complex feature classes are supported.

DETERMINE GEODATABASE REPRESENTATION

ArcInfo lets you store discrete entities as simple features, complex features, and objects.

If the spatial type is point:

- For an unconnected point, such as a historical monument, enter a point feature.

- For a connected point, such as an intersection connected to street segments, enter a simple junction feature.

- For a connected point that has an internal topology, such as a treatment plant, enter a complex junction.

If the spatial type is line:

- For a stand-alone line, such as a fence, enter a line feature.

- For a linear feature that participates in a system such as a road network, enter a simple edge feature.

- For a linear feature with connected sections, such as a section of utility line, enter a complex edge feature.

If the spatial type is area:

- For a stand-alone area, such as a park, enter a polygon feature.

- For space-filling areas, such as vegetation cover, enter a polygon feature (later assigned to a planar topology).

If the spatial type is image (photograph, scanned map, satellite image, or other), enter a raster.

If the spatial type is surface:

- For surfaces in which terrain detail is important, enter a TIN.

- For surfaces covering large areas and to utilize existing digital elevation models, enter a raster.

If the spatial type is an object, enter an object. These are entities that do not have direct geographic representation, but are related to geographic features.

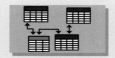

4   Match to geodatabase elements

apply feature geometry
and topology

entity	related to	spatial type	ArcInfo type
Water utility			
Pump	–	point	object
Meter	–	point	point feature
Meter box	Meter	point	point feature
Valve	–	point	simple junction
Water main	–	line	complex edge
Treatment plant	–	point	complex junction
Land records			
Parcel	–	area	polygon feature
Easement	–	line	line feature
Parcel description	Parcel	text	annotation feature
Parcel photograph	–	image	raster
Owner	Parcel	object	object
Address	–	location	address
Streets			
Street	–	line	line feature
Bridge	–	point	point feature
Name	Street	text	annotation feature
Traffic light	–	point	point feature
Bus route	–	line	line feature
Bus stop	–	point	point feature
Environment			
Historic monument	–	point	point feature
Fence	–	line	line feature
Vegetation cover	–	area	polygon feature
Place names	–	text	annotation feature
River valley	–	surface	TIN
Satellite image	–	image	raster

Determine feature and geometry type

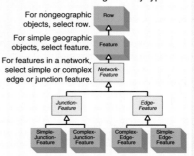

For nongeographic objects, select row.

For simple geographic objects, select feature.

For features in a network, select simple or complex edge or junction feature.

Specify topological graphs

For linear systems, such as transportation or utility, select geometric network.

A geometric network has custom behavior built in to make the editing of networks easy.

For systems of land or jurisdictions, a planar topology manages the shared geometry of a set of features.

A planar topology enforces that no feature can cross another without an intersection.

Implement attribute types for objects

Each entity can have many attributes. These are the attribute types.

short integer	**whole numbers**
long integer	*143* *64* *-14* *56*
float	**real values**
double	*10.0* *2.3* *-4.7* *8.63*
text	**descriptions** *purple mountains*
date	**time** *6 September 1999, 8:20*
objectID	**identifier** *239648547593*
BLOB	**multimedia** *flightOver.mov*

The objective of this step is to identify and name the geographic datasets that will contain the various entities, and in the case of the coverage dataset, to organize entities into coverages.

In this step, you will

- Assign entities to feature classes and subtypes.

- Group related sets of features into geometric networks or planar topologies.

- Organize feature classes and datasets into geodatabases.

Grouping feature classes

In the previous step, you assigned feature types and attributes to entities. Now, you will define the structure of feature classes with subtypes and whether they stand as separate feature classes or are contained within a feature dataset.

Your first consideration is whether an entity should be mapped to a subtype or an entire feature class. Your preference should be to consolidate related entities as subtypes within a feature class, because fewer feature classes will yield better-performing geodatabases. Here are the circumstances when it is necessary to create new feature classes instead:

- When each group of related features requires distinct custom behavior.

- When the set of feature attributes is substantially different. (All features in a feature class have the same set of attributes.)

- When you require distinct access privileges for each group of features.

- When some features are to be accessed through versions and some are not.

Defining topological roles for feature classes

You have defined the feature types for entities.

If the feature type is simple edge, simple junction, complex edge, or complex junction, then it participates within a geometric network. All the feature classes for a geometric network must be placed within a feature dataset. This enforces that they share a common spatial reference.

If the entity feature type is line or polygon and the entity is either meant to cover a complete area, such as land parcels, or if you wish to enforce that crossing features have intersections, then place those features within a common feature dataset. In the ArcMap Editor, you can perform topological editing on these feature classes. This assemblage is called a planar topology.

For entities with simple features, you can also place them within a feature dataset, which also serves as a container for you to arbitrarily group feature classes that are similar.

Gathering datasets and feature classes

Once you have defined your set of feature classes and their topological associations, it is time to group them into geodatabases.

These are some considerations for grouping feature classes and feature datasets into distinct geodatabases:

- If you are working in a large organization, different departments have responsibility for various datasets. Geodatabases can be deployed to follow your organizational structure.

- You have the freedom to use any number of commercial relationship databases, but each must be served through a separate geodatabase.

- If you are working with personal geodatabases, practical size limits may require thematic or spatial partitioning of geodatabases.

Geodatabase
Feature dataset
Feature class
Geometric network

Organize geodatabase structure

entity	related to	spatial type	ArcInfo type
Water utility			
Pump	–	point	object
Meter	–	point	point feature
Meter box	Meter	point	point feature
Valve	–	point	simple junction
Water main	–	line	complex edge
Treatment plant	–	point	complex junction
Land records			
Parcel	–	area	polygon feature
Easement	–	line	line feature
Parcel description	Parcel	text	annotation feature
Parcel photograph	–	image	raster
Owner	Parcel	object	object
Address	–	location	address
Streets			
Street	–	line	polyline feature
Bridge	–	point	point feature
Name	Street	text	annotation feature
Traffic light	–	point	point feature
Bus route	–	line	line feature
Bus stop	–	point	point feature
Environment			
Historic monument	–	point	point feature
Fence	–	line	line feature
Vegetation cover	–	area	polygon feature
Place names	–	text	annotation feature
River valley	–	surface	TIN
Satellite image	–	image	raster

	WaterSystem
geodatabase	
feature dataset	WaterFeatures
object class	Pump
point feature class	Meter
point feature class	MeterBox
geometric network	WaterNetwork
simple junction feature class	Valve
complex edge feature class	WaterMain
complex junction feature class	TreatmentPlant

	Land base
geodatabase	
feature dataset	Land parcels
planar topology	Subdivision
polygon feature class	Parcel
line feature class	Easement
annotation feature class	Description
object class	Owner
relationship class	Ownership
raster dataset and rasters	Lot images — Image
locator and address	US Postal — Address

	Streets
feature dataset	Streets
line feature class	Street
polygon feature class	Bridge
line feature class	Name
point feature class	Traffic light
line feaure class	Bus route
point feature class	Bus stop

	Environment
feature dataset	Environment
point feature class	Monument
line feature class	Fence
polygon feature class	Vegetation
annotation feature class	Names
TIN dataset	Valley
raster dataset and rasters	Landsat — Images

Index

Books from ESRI Press

Modeling Our World
With this comprehensive guide and reference to GIS data modeling and to the new geodatabase model introduced with ArcInfo 8, you'll learn how to make the right decisions about modeling data, from database design and data capture to spatial analysis and visual presentation. ISBN 1-879102-62-5

GIS for Landscape Architects
From Karen Hanna, noted landscape architect and GIS pioneer, comes *GIS for Landscape Architects*. Through actual examples, you'll learn how landscape architects, land planners, and designers now rely on GIS to create visual frameworks within which spatial data and information are gathered, interpreted, manipulated, and shared. ISBN 1-879102-64-1

The ESRI Guide to GIS Analysis
By the author of the best-selling GIS classic *Zeroing In: GIS at Work in the Community* comes an important new book about how to do real analysis with a geographic information system. *The ESRI Guide to GIS Analysis, Volume 1: Geographic Patterns and Relationships* focuses on six of the most common geographic analysis tasks. ISBN 1-879102-06-4

Extending ArcView GIS
This sequel to the award-winning *Getting to Know ArcView GIS* is written for those who understand basic GIS concepts and are ready to extend the analytical power of the core ArcView GIS software. The book consists of short conceptual overviews followed by detailed exercises framed in the context of real problems. ISBN 1-879102-05-6

GIS for Everyone
Now everyone can create smart maps for school, work, home, or community action using a personal computer. Includes the ArcExplorer™ geographic data viewer and more than 500 megabytes of geographic data. ISBN 1-879102-49-8

Transportation GIS
From monitoring rail systems and airplane noise levels, to making bus routes more efficient and improving roads, this book describes how geographic information systems have emerged as the tool of choice for transportation planners. ISBN 1-879102-47-1

Enterprise GIS for Energy Companies
A volume of case studies showing how electric and gas utilities use geographic information systems to manage their facilities more cost effectively, find new market opportunities, and better serve their customers. ISBN 1-879102-48-X

Managing Natural Resources with GIS
Find out how GIS technology helps people design solutions to such pressing challenges as wildfires, urban blight, air and water degradation, species endangerment, disaster mitigation, coastline erosion, and public education. The experiences of public and private organizations provide real-world examples. ISBN 1-879102-53-6

More ESRI Press titles are listed on the next page ➢

*ESRI Press publishes books about the science, application, and technology of GIS. Ask for these titles at your local bookstore or order by calling **1-800-447-9778**. You can also shop online at **www.esri.com/gisstore**. Outside the United States, contact your local ESRI distributor.*

ESRI Press ■ 380 New York Street ■ Redlands, California 92373-8100

Books from ESRI Press

Serving Maps on the Internet

Take an insider's look at how today's forward-thinking organizations distribute map-based information via the Internet. Case studies cover a range of applications for Internet Map Server technology from ESRI. This book should interest anyone who wants to publish geospatial data on the World Wide Web. ISBN 1-879102-52-8

Zeroing In: Geographic Information Systems at Work in the Community

In twelve "tales from the digital map age," this book shows how people use GIS in their daily jobs. An accessible and engaging introduction to GIS for anyone who deals with geographic information. ISBN 1-879102-50-1

ArcView GIS Means Business

Written for business professionals, this book is a behind-the-scenes look at how some of America's most successful companies have used desktop GIS technology. The book is loaded with full-color illustrations and comes with a trial copy of ArcView GIS software and a GIS tutorial. ISBN 1-879102-51-X

Getting to Know ArcView GIS

A colorful, nontechnical introduction to GIS technology and ArcView GIS software, this workbook comes with a working ArcView GIS demonstration copy. Follow the book's scenario-based exercises or work through them using the CD and learn how to do your own ArcView GIS project. ISBN 1-879102-46-3

ARC Macro Language: Developing Menus and Macros with AML

ARC Macro Language (AML) software gives you the power to tailor workstation ARC/INFO® software's geoprocessing operations to specific applications. This workbook teaches AML in the context of accomplishing practical workstation ARC/INFO tasks, and presents both basic and advanced techniques. ISBN 1-879102-18-8

Understanding GIS: The ARC/INFO Method (workstation ARC/INFO)

A hands-on introduction to geographic information system technology. Designed primarily for beginners, this classic text guides readers through a complete GIS project in ten easy-to-follow lessons. ISBN 1-879102-00-5

ESRI Press publishes books about the science, application, and technology of GIS. Ask for these titles at your local bookstore or order by calling 1-800-447-9778. You can also shop online at www.esri.com/gisstore. Outside the United States, contact your local ESRI distributor.

ESRI Press ■ 380 New York Street ■ Redlands, California 92373-8100